INFINITE POTENTIAL

Other Books by F. David Peat

Glimpsing Reality: Ideas in Physics and the Link to Biology, with Paul Buckley

Rituals of Renewal: Spiritual Transformation Through Native American Tradition, with Leroy Bear

Lighting the Seventh Fire: The Spiritual Ways, Healing and Science of the Native American

Todd Watts: New Lamps for Old

Philosopher's Stone: Chaos, Synchronicity and the Hidden Order of the World

Quantum Implications: Essays in Honour of David Bohm, edited with Basil Hiley

Einstein's Moon: Bell's Theorem and the Curious Quest for Quantum Reality

Turbulent Mirror: An Illustrated Guide to Chaos Theory and the Science of Wholeness, with John Briggs

Superstrings and the Search for the Theory of Everything

Synchronicity: The Bridge Between Matter and Mind

Science, Order and Creativity, with David Bohm

Artificial Intelligence: How Machines Think

Looking Glass Universe, with John Briggs

In Search of Nicola Tesla

Infinite Potential

The Life and Times of DAVID BOHM

F. David Peat

HELIX BOOKS

ADDISON-WESLEY PUBLISHING COMPANY, INC.
Reading, Massachusetts Menlo Park, California New York
Don Mills, Ontario Harlow, England Amsterdam Bonn
Sydney Singapore Tokyo Madrid San Juan
Paris Seoul Milan Mexico City Taipei

Many of the designations used by manufacturers and sellers to distinguish their products are claimed as trademarks. Where those designations appear in this book and Addison-Wesley was aware of a trademark claim, the designations have been printed in initial capital letters.

Grateful acknowledgment is made to Charles Biederman for permission to reprint from his correspondence with David Bohm, excerpted in chapter 12; and to the Albert Einstein Archives, the Hebrew University of Jerusalem, Israel, for permission to reprint from Einstein's correspondence with David Bohm, excerpted in chapters 8 and 9.

The photographs in the text are courtesy of Sarah Bohm unless otherwise credited.

Library of Congress Cataloging-in-Publication Data

Peat, F. David, 1938–
Infinite potential : the life and times of David Bohm / F. David Peat.
p. cm.
Includes bibliographical references and index.
ISBN 0-201-40635-7
1. Bohm, David. 2. Physicists—United States—Biography. I. Title.
QC16.B627P43 1996
530′.092—dc20
[B] 96-24420
CIP

Jacket design by Suzanne Heiser
Jacket photograph © by Mark Edwards/Still Pictures
Text design by Dede Cummings
Set in 11-point Sabon by Pagesetters, Inc.

1 2 3 4 5 6 7 8 9 10-MA-0099989796
First printing, October 1996

Contents

ACKNOWLEDGMENTS

I WOULD LIKE to thank David Bohm's friends, colleagues, and relations who have been so generous in assisting me with this biography. By sacrificing their time to grant interviews, engage in correspondence, reply to telephone calls, and search for documents and letters, they have enriched my knowledge and understanding of Bohm's life, the better to recount it. So many helped me that if I were to attempt to name them all, I might risk hurting someone through accidental omission.

I do, however, give special thanks to David Bohm's colleague Basil Hiley, who made a detailed scientific critique of a draft of this book. His insights, as well as his personal knowledge of events, have proved invaluable. He cannot be held responsible for any errors that may have crept into the final draft.

I also want to acknowledge the role played by David Bohm's widow, Saral Bohm. From the beginning Saral agreed that I should have total freedom to write about her husband's life as I saw it. She generously provided access to research material, paved the way for interviews with friends and relations, and talked to me about her life with Bohm. While aspects of this biography may have pained her, she always made it clear that the final decisions always rested with me and that the truth, as I saw it, should be told.

I would also like to thank the Canada Council for the Arts, the Fetzer Institute, and Georg Wikman of the Swedish Herbal Institute for their support.

Infinite Potential

Introduction

On the afternoon of October 27, 1992, David Bohm was at Birkbeck College, the University of London, putting the finishing touches on a book that would sum up his lifelong struggle to create an alternative quantum theory. At six-fifteen he telephoned his wife, Saral, to let her know he was about to leave. "You know, it's tantalizing," he said. "I feel I'm on the edge of something."[1] An hour later, just as his taxi pulled up outside his home, Bohm suffered a massive heart attack and died.

Those last words, with their sense of bordering on the unknown, exemplify the thrust of Bohm's life. The man Einstein once spoke of as his intellectual successor[2] was always seeking to go beyond, to transcend, to ask that next question. He had the courage to pursue truth no matter where it took him, yet he was guided by a strong moral sense. Still, there were many paradoxes to Bohm's life. Once a confirmed Marxist who scrupulously avoided any taint of mysticism, he later devoted much of his time and energy to the Indian philosopher and teacher Jiddu Krishnamurti. Once inspired by the American Dream, he later stood trial for contempt of the U.S. Congress. At Princeton, the same year he seemed set for a Nobel prize, he was suddenly assailed as a "juvenile deviationist" of science whose work should be ignored by the scientific establishment.

Nonetheless, his scientific achievements were more than enough to assure his reputation. In California, during the war years, he developed the theory of the plasma—the fourth state of matter, in addition to the solid, liquid, and gaseous states. At Princeton he applied this theory to the way electrons move in a metal and set the stage for much of subsequent solid-state physics.

His textbook on quantum theory, written while still at Princeton, became a classic for its clarity, always relying on physical argument and philosophical principles to explain the quantum world, rather than falling back on abstract mathematical formulae.

Later, at Bristol University in England, Bohm and his student Yakir Aharonov demonstrated a new and important way in which the quantum world transcends that of classical mechanics. The two physicists showed that an electron is affected by the presence of an electrical field even when, according to classical physics, it is totally shielded from that field. This effect, they argued, is central to quantum mechanics, implying that even quite distant objects can affect quantum processes. These nonlocal correlations have nothing to do with traditional forms of interaction (such as by fields or the exchange of particles); rather, they demand new concepts that go beyond the ideas of separation and distance. The prestigious scientific journal *Nature* editorialized that Aharanov and Bohm's work was worthy of a Nobel prize.

Bohm had also reformulated the paradox proposed by Einstein, Boris Podolsky, and Nathan Rosen (EPR) that attempted to retain "independent elements of reality" within the quantum world. In Bohm's version the meaning of this paradox became clearer and helped blaze the trail for what would later become an experimental test.

This experimental test was proposed by physicist John Bell in his famous theorem. But Bell himself had been led to develop this theorem after encountering Bohm's hidden variables version of quantum theory—in Bohm's 1952 papers, Bell later said, he had seen "the impossible done."[3]

Throughout the later decades of his life, Bohm sought a new order in physics. He proposed that the reality we see about us (the explicate order) is no more than the surface appearance of something far deeper (the implicate order). According to Bohm, the ground of the cosmos is not elementary particles but pure process, a flowing movement of the whole. Within this implicate order, Bohm believed, one could resolve the Cartesian split between mind and matter, or between brain and consciousness.

Bohm's notion of an implicate order extended his reputation outside the bounds of physics and drew the interest of writers, artists, psychologists, and philosophers. It was to this audience that Bohm directed much of his later work, lecturing and writing on the essential wholeness of nature and experience, deploring the fragmentation of our modern world, discussing the nature of creativity, and exploring the nature of thought and the structuring processes of the psyche.

So deeply have his ideas permeated the general culture that they are becoming part of the shared way we look at the world. Their influence can be found in areas as diverse as education, psychology, art, and literary criticism, appearing even in novels. Bohm became something of a guru to those seeking renewal through education and psychotherapy, or seeking to build new communities or understand the internal dynamics of society.

In spite of his considerable scientific reputation, Bohm did not always see eye to eye with his contemporaries. The major controversy of his life lay in his rejection of the conventional interpretation of quantum theory. After his contact with Einstein, Bohm proposed an alternative theory in which electrons are guided along paths by what he called the *quantum potential*. This "hidden variable" theory so offended the scientific establishment that it was met with not only rejection but sheer silence, which gave Bohm considerable pain. Although he went on to develop the theory further, moving away from strict determinism into something far more subtle, his work remained tainted as that of a scientific maverick.

Throughout the 1970s and 1980s Bohm went ever deeper into the quantum theory, seeking to develop a theory of prespace that would make connections to Einstein's relativity. It was during this period that Bohm moved away from his earlier materialistic position. Attempting to remove the distinction between mind and matter, he proposed that information, like matter and energy, is one of the basic principles of nature; it is not a subjective assessment but an objective activity in the world. The more broadly his ideas ranged, however, the more rigid and hidebound the scientific community became.

At the end of his life, Bohm remained a scientific rebel. He rejected the current fashion of seeking closure in some "grand unified theory," in favor of a vision of nature's inexhaustibility, of a world of infinite levels. Bohm's world was holistic, as holistic as the unanalyzable interconnections of the quantum or his unified vision of matter and mind. Holism extended, he believed, into human psychology and society itself. He dreamed of developing a group mind and spent his last years organizing dialogue circles in its pursuit.

Bohm lived for the transcendental; his dreams were of the light that penetrates. From early childhood he learned to escape into the world of the mind and the imagination. Yet his life was accompanied by great personal pain and periods of crippling depression. He never achieved wholeness in his own personal life, and the fruits of that life, which are still with us, were gained only at great sacrifice.

CHAPTER 1

Childhood: From Fragmentation to Flow

IN HIS LATE MIDDLE AGE, with his tweed jacket and shy hesitant manner, David Bohm had all the appearance of an English academic. But he was actually born on December 20, 1917, in the Pennsylvania mining town of Wilkes-Barre. His nationality was American, but his family origins lay in Eastern Europe. His father, Shmuel (known to the family as Shalom), had been born in the Hungarian town of Munkács, a thriving business center of the Austro-Hungarian empire founded in the tenth century. It was the first major city that emigrating Jews reached after they had crossed the Carpathian mountains via the Veretsky Pass, and in consequence, it had a large Jewish population. The Chasidim of Munkács were reputed to be particularly learned. Following World War I the town became part of Czechoslovakia and was renamed Mukačevo. In 1945 it was absorbed into the Ukrainian region of the USSR.*

Shmuel's family name at that time was not Bohm but Düm. Shmuel's father, Aaron David Düm, had married Esther Kalish. Their nine children, Shlomo, Yenta, Moshe, Leah, Shmuel (David Bohm's father), Raisel, Ezekial, Mirel, and Yaakov, were born into an Ortho-

* Some of the information on the town of Munkács is from Leslie Ivan, a neurosurgeon living in Ottawa. He and Bohm met during one of the physicist's visits to Ottawa and found that they had much in common, including a Hungarian background.

dox Chasidic home. In 1904, at the age of thirty-five, Aaron died, and within the year Esther was also dead. (David Bohm believed that the cause of his grandfather's death was cholera; his cousin Irving thought it was influenza.) Tragedy haunted the family, and during the Second World War most of the Düms—David Bohm's uncles, aunts, and cousins—perished in the Holocaust.*

Their parents dead, the nine Düm children were raised by Jewish families in the town. Shmuel benefited from a good education at the yeshiva (Jewish academy of learning), where he studied the Talmud. An intelligent boy, he would have preferred to continue with his studies rather than be thrust, as he was, into the world of business. Religion remained important to him, and as an adult he served as a shamus—an assistant to a rabbi—with the authority to perform circumcisions, slaughter animals, and supervise the preparation of food. Noted for his love of opera and theater, he had a good voice and enjoyed singing in the temple: later he became a cantor.

By the time Shmuel reached his late teens, the Munkács families had decided that his future lay in the New World. He was sent to America with a list of names of Jewish families who would look after him. Upon landing at Ellis Island, he was told by an immigration official that his name, *Düm,* would mean "stupid" in English. The official changed the name to Bohm. As he established himself in America, Shmuel resorted to the more familiar Sam, to become Samuel Bohm.

Torn from his hometown and his Talmudic studies, Samuel made his living as a peddler in Pennsylvania. He took lodgings with the Popky family in Wilkes-Barre, a small town populated mainly by Polish and Irish coal miners. It was there that he laid plans for his future.

* The youngest son, Yaakov, was killed in World War I at the age of twenty, fighting with the Hungarian separatists. Yenta died along with two of her children in the Holocaust; one daughter survived. Moshe Düm married Sarah Hollander and, after having two children in Munkács, emigrated to the United States, where four more children were born. Shlomo, too, came to the United States, married, and had two children. Leah married and moved to Rumania, where she, her husband, and six of their eight children perished in the Holocaust. Two daughters survived and went to live in California. Raisel and seven of her children also perished in the Holocaust; one daughter survived and moved to Brooklyn. Ezekial had four children, two of whom died in the Holocaust; two others survived, one to settle in the United States and the other in Israel.[1]

The Popkys were an Orthodox Jewish family from Lithuania. According to the family story, before he emigrated Harry Popky served two terms in the army so that his brother could attend medical school. Arriving in the United States virtually penniless, the story went, Harry stole ten dollars and used the money to buy goods, which he then sold at a profit. From that point on there was no looking back, and eventually he operated a thriving secondhand furniture business. He was said to be worth as much as $100,000—a considerable fortune in those days.

Harry Popky had a non-Jewish friend, Horst Bennish, who owned a large furniture store and obtained secondhand furniture in partial payment for the new furniture he sold. He passed the secondhand furniture on to Harry, who put it up for sale at rock-bottom prices. The Popky's store was in the Polish part of town, and Harry Popky made it his practice to drink schnapps and joke with his customers in Polish. A traditional bargaining session would follow, so that by the time the miners departed with their furniture, they felt not only that they had made a good deal but that honor had been satisfied. Harry, for his part, made a fair profit. It should be added that Harry's fortune did not last indefinitely. He was particularly susceptible to the hard-luck stories of European immigrants, and by the end of his life, he had parted with most of his money.

Good business sense, geniality, and the ability to put his customers at ease were Harry's particular skills, and he spotted a man of the same ilk in Samuel Bohm. Harry also had an ulterior motive in introducing the young man to the furniture business: In the old country it had been the custom for a father to think in terms of a *shiddach* (an arranged marriage) for his daughter. Harry happened to have a daughter, Frieda, who posed particular difficulties in this respect. While the Popky family were still in Europe, Frieda had exhibited a lively intelligence, but on arriving in America and unable to speak the language, she became extremely quiet and withdrawn (a state that eventually developed into more serious mental abnormalities). Possibly to remove any taint of mental disorder, family members claimed that she had been dropped on her head as a baby. At all events, arrangements were made for Samuel to set up a furniture store of his own, complete with a wife.

Samuel and Frieda Bohm found themselves living over their store at 410 Hazel Street. If Samuel was never quite as successful as his father-in-law, at least he was able to provide a comfortable life for his family and help his brothers emigrate to the United States. In these early days he took an interest in socialism and tried to talk to the miners about social principles. Their reaction was laughter; Jews only wanted to make money and had no right to talk about socialism, they said. From that point on, Samuel adopted the persona that was socially expected of him—lively, sociable, and entertaining, with a roving eye for a good-looking woman. His approach to business seems to have been inspired by his father-in-law; his tactic was to overprice his furniture and then allow his customers to beat him down. His customers felt good, he joked, because they believed they had outdone a Jew at business. In many respects, anti-Semitism was alive and well in Wilkes-Barre.[2]

The business prospered, and on December 20, 1917, the Bohms' first child, David Joseph, was born. Four years later David's brother Robert (Bobby) appeared, and the family was complete. But Frieda was becoming increasingly confused and unable to run the house smoothly. That responsibility shifted to her mother, Hanna Popky, who lived nearby. Hanna, a devout woman who knew very little English, became the calm, reliable center of both the Popky and Bohm households. Where Harry might be extravagant, Hanna put aside a little money toward her grandsons' education. As a very small boy, David remembered Hanna and the visits to her home, where Harry Popky would pay him five cents to say prayers in Hebrew. Yiddish was always spoken in the Popky house, but David preferred not to join in, feeling he was not being truly American if he did.

Thanks to Hanna Popky, the Bohm household had a semblance of normality, at least when it came to meals and laundry, but following Hanna's death, things fell apart. Frieda's mental instability left her incapable of coping with the routines of daily life, and it fell to Samuel to feed the boys and get them ready for school each morning. Much of the time Frieda was in a state of depression that could swing unpredictably into agitation and even violence: on one occasion she broke a neighbor's nose with a bottle. There were even periods when she heard voices and experienced other psychotic symptoms, becoming overwhelmed with rage and threatening to kill her husband. David's cousins joked about a roast chicken she had served them—complete with its burnt feathers and innards.

Despite her erratic behavior, David loved his mother and was attracted to what he felt was her spiritual quality and inwardness. Later in life he believed that he had inherited his particular sense of the material world from his mother. She had an intuitive understanding of the world, he felt, but was too confused to act correctly or make sense of the world around her. David inclined toward her emotionally, and she, in turn, made him her favorite. The closeness of their bond made his younger brother, Bobby, jealous.

By contrast, Samuel Bohm seemed brash and materialistic to his elder son, interested only in making money or playing cards. David was disturbed that he constantly belittled his wife, hurling insults at her until she became enraged. Each evening when his father came home from the shop, the tension thickened. Samuel would also direct his sarcasm toward David, his mother's favorite.

David's times alone with his mother could have held intimacy and empathy, yet they too were spoiled by her obsessions. She was always worrying about him, always asking him how he felt. When he grew a little older, she had a neurotic concern every time he went outside. Inevitably, David began to worry about his own health and anticipate the worst. One time he cut his finger and, frightened, left part of the fingernail hanging. His mother panicked to the point that she was unable to help him, which increased his own fear. Finally one of his uncles entered the room and took control of the situation simply by putting David's finger under the tap and then dressing it. At one stroke a problem rendered insurmountable by fear and confusion had been resolved by a clear, practical action.

At times like that David realized how fearful he had become, that he was overly concerned about his health and always worrying about things. If only, he fantasized, he had been brought up in the clean air of the Wild West. Then he would have grown up strong and able to face life like other boys.

When David was seven or eight, the problems in his parents' marriage came to a head, and they considered separating. An impossible choice would be thrust upon him if they did: Where would he go?

He loved his mother and disliked his father's dismissive attitude; probably he even feared him. Yet he knew that his mother was incapable of looking after him, and that he and Bobby would have to live with their father. As it turned out, Samuel and Frieda stayed together, but the quarreling and tensions continued.

At elementary school there was little to suggest that David would amount to anything out of the ordinary. Then a curious quirk of fate directed him toward the path he would pursue for the rest of his life. The year was 1928, and David was ten. One of the boys Samuel had hired to help out in the store left behind a magazine called *Amazing Stories*. David read it and was particularly struck by "Skylark of Space," the story of a rocket journey to distant planets.*

The story fired David's imagination, and he went out and bought *Amazing Stories* and the other science fiction magazines, *Astounding Stories* and *Wonder Stories*, as they appeared. With a school friend, Henry Kunicki, he made up stories and planned a trip to a planet. Henry warned him that the new world would be so dangerous that as soon as they landed there, they might have to flee back to Earth. But David did not want to return to Earth. He decided he would remain on the planet long enough to explore it and contact its advanced civilizations. This worried Henry so much that he threatened to tell David's parents about his plan.

There is nothing particularly unusual about such fantasies. Many children have imaginary friends or travel to secret countries through the back of a cupboard. Even some adults have a "secret garden," a place of retreat from the hurts of the world. But what was exceptional in Bohm's case was the energy with which he invested his imagination. David's space travel had become so vivid that Henry Kunicki honestly worried that his friend would be stranded on a distant, hostile planet. Yet instead of the monsters and hostile aliens typical of pulp science fiction, David's fantasy planet was inhabited by beings who were

* Bohm himself recalled the story a little differently, believing he had been eight at the time and that the story was entitled "Columbus of Space." However *Amazing Stories*, the first science fiction magazine, appeared annually starting in 1926, running reprints of the classical stories of Jules Verne and H. G. Wells. Not until 1928, when it commenced quarterly publication, did it run original stories. "Doc" Smith's "Skylark of Space" was one of the first. It contains all the elements that Bohm later remembered.

scientifically and morally more elevated than his companions on earth. It probably goes without saying that if David, in his imagination, was going to escape from home and Wilkes-Barre, it had to be to a place in his dreams.[3]

Einstein once said that his scientific interest had been aroused by watching the movement of a compass needle that he had been given as a child. Others have been inspired by a teacher at school, by collecting plants or minerals, looking through a microscope, coming across a textbook, or attending a lecture. For David, it was not an empirical fascination with the natural world that fired his passion but science fiction—flights of fancy that took him far from the commonplace and practical into the realms of the imagined possible. His desire to go beyond the security of the familiar remained with him throughout his professional life. He always worked at the edge, accepted nothing at face value, sought the unknown rather than the known, and let himself be guided by intuition and imagination rather than logic and empiricism.

Little in the way of science was taught at the schools he attended, and his first encounter with the field was not through formulae, equations, and categories of facts but in the vast integrating sweeps of limitless space, voyages to the stars, unbounded energies, and the advanced intelligences of pulp magazines. His experience of the more orthodox face of science came only later in his school years, when a series of readers were distributed to his class. He was lucky enough to be given one containing a simplified account of astronomy. While the other children were still plodding along at the pace set by their teacher, David raced to the end of the book. His imagination was fired by the stars and planets, the great distances that lay between them, and the harmonious motion of the solar system. He could not help but contrast the order of the heavens with the chaos of his own home. Everything in his daily life seemed oppressive. Even the teachers at school appeared more interested in exercising authority than in answering his questions. In bed at night he told his younger brother stories about an army of bogeymen—but worse than any of them were the teacher-bogeymen.

Certain ideas and images began to obsess him. It is difficult to know what to call them, for they were more vivid than the pastel shades of daydreams. They have more the graphic nature of true dreams, or of highly charged fantasies that are repeatedly revisited until the energy of active imagination vividly illuminates them. Their subject was often light. David wondered if the light from a streetlamp would reach the stars and beyond. He dreamed of a light of such power that it would penetrate all matter—a light so intense that its color transcended blue and ultraviolet into some unknown color beyond. Later, he dreamed of fingers of light that could reach into and probe his own brain. Energy in all its forms interested him. And so he dreamed of being able to harness lightning and liberate energies so great that they could destroy planets. Later, when he studied chemistry, the active elements excited him, the metals sodium and potassium that ignite when placed in water, and the gas fluorine, so reactive that its vapor will etch glass. He felt stimulated when he read about the effects of adrenaline upon the brain.

Being able to control the limitless power and highly destructive energies within the cosmos of his imagination must have given David a great sense of elation. He read avidly about the power of tornadoes. A field where he played was scattered with large blocks of stone, said to be the remains of a mill destroyed by a tornado. On one occasion a tornado approached Wilkes-Barre; the family huddled indoors for safety. Afterward they discovered that the large plate-glass window of the furniture store had been destroyed. The boy who had been working in the store told them how the entire window had blown in toward him until, at the last moment, it flew out into the street. Surveying the damage, David noticed that the cast-iron roof of a water tank had blown off and landed on the roof of a house. Later, he tried to recreate such twisting masses of air in the kitchen, holding a can of ice water over the hot stove.

One does not have to look far for the origin of David's fantasies. All children need love, physical closeness, encouragement, and emotional stability. They must learn to trust the outer world of objects and human relationships if they are to journey from the amorphous buzzing confusion of infancy into the differentiation of self from other and the discrimination of thoughts, emotions, and sensations. Few of us grow up within an ideal environment; nevertheless, human development has an inevitability about it. Nature programs us along a path that is sufficiently robust that most of us can slough off the traumas and deficiencies that the contingent world forces upon us. But there are always some individuals who are more sensitive and more finely adjusted than the majority.

The home into which David was born was chaotic, oppressive and at times, violent. His first years were spent in fear that these external tempests would overwhelm him, flooding inward and drowning his tenuous sense of self. In a child's fight for emotional survival, the impotent paralysis of fear finally gives way to a rage so great that it threatens to crack the universe and destroy the very beings on whom the child depends. In our rage, we pluck planets from the sky, burn suns, rip the very fabric of space and time with our fingers. But rage brings with it guilt, the fear of losing control and the consequent dread of reprisal.

These are the conventional psychological explanations for David's dreams of light, power, control, and transcendence. Yet David was also possessed of a special genius that, as it unfolded throughout his life, would make his vision richer and more formal than any mere flight from reality. Had he lived in an earlier age, he might have spoken of communicating with gods or making compacts with the energies of the universe. Our own time prefers the psychological language of a fragile ego inflated by the potential of the (Jungian) Self. Whatever the metaphor, such powerful psychological material results in the person becoming literally "inspired" and "enthused." Throughout his life Bohm yearned for contact with the transcendent and for moments of "breaking through." Yet for the theoretical physicist, as much as for the shaman, artist, poet, or musician, the personal cost of such yearning is very great. Only the very strongest are able to contain the powers that animate the African mask, the Greek tragedy, the Dionysian revel, and the infinite vacuum state of theoretical physics. In his personal life David paid the price for access to such powerful forces. Through the medium of his own suffering, he created work that inspired and illuminated many others.

David's other place of retreat was nature: the woods and hills that surrounded the town, where he loved to wander alone and dream. But even here things were not ideal, for David suffered from physical awkwardness and a lack of coordination. While other boys played sports with ease, David could not even learn to catch a ball and would sit out games on the fringes, watching others play.

A lack of physical coordination can place a child socially beyond the pale. Incapable of making spontaneous movements and expressing their enthusiasm for life, they become the butt of cruel jokes and are excluded from group adventures. The cause of their problem may be a defect in perception, a lack of muscular coordination, or an inability to experience the body correctly in space. But it can also be the symptom of a psychological failure to engage fully with the external world. Closing the body and armoring it to the world can be an effective defense against external hurts. Whatever its cause, its effect is often an intense sense of alienation.

David certainly wanted to move and play like other boys, and around the age of seven or eight, he decided that by watching what they did, he should be able to work out in his head the various bodily movements involved in catching a ball or climbing a tree. By carefully planning each move, he would then be able to control his body. A few years later, at around ten or twelve, he was walking in the woods with a group of boys when they came to a stream traversed by a series of rocks. Again, it was the sort of situation that troubled him. He would now have to plan ahead, note the position of the rocks, and decide where and how to place his feet. For David, physical security came in assuming trusted positions: he would move only when he had developed sufficient confidence. Yet as soon as he jumped onto the first stone, he realized that it was impossible to stop long enough to plan the next step. Crossing the river, jumping from stone to stone, could be done only in one continuous movement. If he tried to stop or even think about what he was doing, he would fall in. His only hope was to keep moving.

This moment of insight became so significant to him that he told the story many times during his life. Up to that point, David had assessed each situation in his life, never fully committing himself, always fearful of being pulled along by "irrational currents."[4] At that moment, however, he suddenly realized that security does not require control and stillness but can come in a freely flowing movement.

It would be satisfying to report that from that day onward, David was determined to enter college on a football scholarship or join a *corps de ballet*. But his outer awkwardness remained. The first impression people had of him was of a brilliant mind far from ease within his body. Decades after he crossed the stream, he attended a social occasion where he was persuaded to dance. But before he could proceed, he planned each foot position in advance, with reference to the tiles on the dance floor!

Still, a deep transformation had occurred at an inner level. David began to see the world in terms of flows and transformations, processes and movements. No longer would he think of the world exclusively in terms of fixed objects like stones and atoms, for behind them he now saw a previously hidden world of motion and flow. This world connected with his other preoccupations. In the bathtub he noticed that as the water spun down the drain, it looked like a miniature tornado. Each time he tried to disturb it, the vortex formed again. Something stable was emerging out of movement itself. Take away the movement of the water, and the vortex would vanish. Again it occurred to him that stability and existence could be created out of pure process.

Around the time of bar mitzvah, when according to Jewish tradition, a boy becomes a man, David began to share his enthusiasm for science fiction with a boy called Mort Weiss. Mort was the son of a local salesman, Samuel Weiss, who lived a mile or so away in the Jewish part of town. Samuel Weiss was a good friend of Samuel Bohm—at the temple on the Sabbath, the two men would hold religious discussions that would continue into one of their homes. It was natural that Samuel Weiss, a man learned in the Torah, should instruct his son for his bar mitzvah. Mort and David went through the ceremony together, attended the same school, and became good friends.

The two boys did their homework on the Bohms' kitchen table. Frieda Bohm tried to be kind, sometimes giving the boys hot cocoa and cupcakes. Yet Mort noticed that she dampened the atmosphere and that David seemed to ignore her. When David's father returned home from the shop, the tension would increase, at which point Mort would leave.[5]

The boys were not supposed to work on the Sabbath, but they enjoyed visiting the local library on Franklin Street, where David surrounded himself with piles of books. He took science fiction seriously enough to send letters to magazine editors: "When I think," he wrote, "of the real science-fiction we were getting two and three years ago (that is how long I have been reading 'the magazine'), it seems that 'Wonder Stories' has been for a long time degenerating into one of the mediocre pulp magazines with which the newsstands now abound." His objection was that the current stories were "adventure or love stories with a sprinkling of scientific background," clearly aimed at increasing circulation. Possibly, David suggested, the authors were getting old or needed a vacation. "But whatever the cause, science-fiction is doomed unless new super-plots are developed with the same technique as the older stories had." He proposed that a premium be offered for well-written stories.

While David enjoyed reading about aeons of history and light-years of space, it always had to be based on strict scientific principles. He fired off another letter to complain about a story concerning Drusonians, whose eyes could see no blue. To make themselves invisible, humans painted themselves blue, but as David pointed out, the effect would be exactly the same as if we saw people covered in ultraviolet paint—they would not be invisible but would appear black. "Black is not the least bit invisible," he pointed out. Raising this point was not hairsplitting, he argued, since the entire story was based upon that erroneous assumption.[6]

When they weren't doing homework or reading science fiction together, David and Mort built crystal sets. In those days radios were simple; they required no power source and consisted of a wire antenna, a crystal detector, a coil, and headphones. A galena crystal acted as a rectifier for the oscillating current picked up in the antenna. The boys would jiggle a thin wire called a cat's whisker across one of the crystal's faces until they could hear a strong signal.

The local radio station was experimenting with broadcasting, and the boys felt that they were living at the beginning of a new age. David enjoyed designing circuits and sorted through the dumps of coal tailings behind his home, looking for bright-looking crystals. Fascinated by the future, he told Mort about television and about aircraft that would fly without propellers, and how spacemen would have to carry their own oxygen when they visited the moon.[7]

Samuel Bohm thought little of David's passion for "scientism," as he called it. It was all "up in the clouds," and he scorned anything that was not practical. Bobby, who loved sports, was closer to Samuel's idea of what a boy should be. In exasperation, Samuel insisted that David take more of an interest in boyish pursuits and enrolled him in the local Young Men's Hebrew Association. But the meetings of the sports club bored David, who spent most of the meeting times with his nose in a science book. In the hopes of whetting his athletic interest, one of the adults at the club gave him a sports magazine for boys. Leafing through it, David found a story about a boy who threw a ball so hard that it entered the fourth dimension, only to return inside out. The idea of a fourth dimension fascinated him. Maybe the world around him was less substantial than everyone believed. Could the world be just the projection of some higher dimensional reality? Later, when he was learning about solid geometry in school, he worked out the consequences of living in a four-dimensional universe, in what was essentially his first piece of speculative physics.[8]

While David dreamed of space travel and the fourth dimension, Samuel's reverie was that his sons would one day run the largest furniture store in Wilkes-Barre. He was to be badly disappointed. David rejected what he felt was his father's undue concern for money and the purely practical side of life. Yet in some ways David's image of his father was distorted. Samuel loved music and theater; he was intelligent, and he was religious to the point of assisting the local rabbi. Clearly there was more to him than making money. David, however, remembered only his father's ignorance, his failure to realize that the planets were other worlds that orbited the sun. When David tried to explain it to him, his father replied that none of it really mattered since it had little to do with human relationships and daily social life.

David's negative opinion of his father was not shared by his younger cousins, who lived in Philadelphia. They eagerly anticipated Uncle Samuel's visits. He always brought presents for Ruth, his niece, and would play games and sing to her. How modern he seemed compared with the formality of her own father! Rumor had it that Uncle Samuel's wife was unable to look after him, so Shlomo's family would always make sure that clean shirts were available for him.

Uncle Samuel loved to socialize and entertain. By contrast, cousin David was shy, withdrawn, and thoughtful during family visits. He seemed to sleep a great deal. Sometimes of an afternoon his cousins would creep into his bedroom to watch the way he slept with a pillow over his head. Yet despite his reserved nature, Ruth developed a bond with David. She could not help noticing, however, the way his body would stiffen and pull away when anyone attempted to show him affection. She came to think of him as a spirit come to earth, that had been forced to inhabit a physical form in which it never felt at peace.[9]

When Samuel paid visits to his other brother in Philadelphia, Schlomo's son Edward was given the job of entertaining David and Bobby. That was simple when it came to Bobby, who liked to play baseball, but David showed no interest in sports or games. In the end, Edward took him to the nearby Franklin Institute, where he could browse around the scientific exhibits, and left him there for the day.[10] It is something of an irony that this same institution awarded him the prestigious Franklin Medal toward the end of his life.

While Edward found his uncle Samuel outgoing, he also recognized that Samuel was not the sort of man to whom a young nephew could get close.

David's own feelings towards his father were confused. Although he felt anger toward the man who rejected everything he was interested in, it was natural that he should also want to win his approval. He had to show his father that science could be useful and that ordinary people could understand even the most difficult scientific ideas.

In later life David was fairly hopeless when it came to practical matters, but while he was living in his parents' home, he attempted to make a series of inventions. One of these was a dripless pitcher, which he worked on to the point of making an affidavit of its invention. He even compiled a list of companies that manufactured or used bottles, who might be interested in the pitcher. He read an advertisement that, upon payment of one dollar, offered advice on how to market an invention. David mailed off his dollar and was told to sell his pitcher door to door.*

* "Duplicate of Affidavit sworn on Sept 12 1933. Sworn before Charles P (Illegible). Witnessed by Frances Sheperd. I, David Bohm of 410 Hazel St, Wilkes-Barre Pennsylvania, United States of America do hereby make an affidavit stating that I have invented an improvement on pitchers, bottles, and other similar containers of liquids, of which the following is a specification: . . ."

He made other inventions, too, on paper at least. On January 22, 1934, he recorded his invention of the All Wave Superheterodyne:

"The purpose of this invention is to make possible the construction of a radio receiver which can tune all waves from 9 meters to 1000 meters and over on one dial, without switches, plug in coils, and any such devices." He went on to say that the circuit would eliminate "image frequency interference."

David's Vertical Wing Tip was designed to increase lift by preventing air from leaking over a wing's edges: "the ideal wing is one that is short, broad, and of rectangular shape, but which would yet possess a very small air leakage over its tip. A wing of this type has not yet been constructed; but on this day of Feb 23, 1933 A.D. I, David Bohm, have invented a device which will accomplish this purpose."[11]

A further advance was the Two Stroke Cycle Compression-Ignition Engine, designed "to improve the Diesel engine in such a way that it might operate with one power stroke per revolution instead of the usual power stroke per two revolutions. A hollow piston is used with explosions taking place within and above the piston."

David's inventions never reached the marketplace, but his belief in the importance of science for everyday life remained with him. Throughout his life he tried to make his scientific ideas, no matter how abstract, accessible to ordinary people. In addition, at least until he moved to Europe at the end of the 1950s, he believed that science would be instrumental in transforming society. His father, who had once looked down on David's "scientism," became proud of his son when his first book, *Quantum Theory*, appeared, and he attended the Princeton party in honor of the young physicist.

In Wilkes-Barre the Bohms lived a mile or so from the Jewish community, whose businessmen, David noted, tended to look down

---

He explains that liquids, when poured from a container, tend to follow the outer surface of the container rather than falling directly. His invention was a collar that "can safely sustain an edge sharp enough to break the film of surface tension." The collar would project from the rim to give a greater angle of cant without increasing the flow of water. Further, he bent the collar outward, "so as to allow the liquid to gain velocity before reaching the rim and also to cause gravity to pull against surface tension on the lower surface. . . . In all cases I have tried it, no liquid has dripped when poured over this new type of rim." On another sheet of paper Bohm listed "Addresses of Milk Bottle companies" obtained from the Purvin Dairy Company.

on Poles as poor and lacking in intelligence. For their part, the Polish families regarded Jews as standoffish and having too high an opinion of themselves. Neither bias made for ethnic harmony, and the miners' children usually adopted their parents' prejudices. On Sunday mornings, Mort Weiss remembered, the Polish boys would emerge from church, "all whipped up and angry," looking for Jewish boys to beat up. Because his elder brother was a tough fighter, Mort got some protection. The same protection did not, however, apply to David, who was the only Jewish boy in his neighborhood.[12]

Looking back with the hindsight of half a century, David Bohm was philosophical and did not recall particular examples of bullying. He did, however, remember being called a "dirty Jew" and accused of crucifying Christ. He responded to his tormentors that it was the Romans who had crucified Christ, and anyway he couldn't be held responsible for what had happened two thousand years ago. All in all Wilkes-Barre was a rough environment in which to grow up. David remembered the local gangs, such as the Mayflower Gang and the Blackman Street Gang, who spent their time hunting each other down or looking for boys to beat up.[13]

It may be difficult to believe that this delicate and uncoordinated Jewish boy survived his school days without trauma. Yet from the security of old age, David felt that he had gotten along well with the local boys. At first, the taunts he did receive made him extremely angry, but the other boys, and their parents, advised him not to answer back or otherwise show that he had been irritated. What did strike him, however, was his father's response of backing down when faced with the possibility of physical aggression. David felt a sense of betrayal when the father who had belittled him for his own physique nonetheless failed to stand up for himself (not an unreasonable reaction for a furniture salesman confronting an angry miner!).

On one occasion an Irish boy so taunted David that he told the boy's parents, who replied that there was little they could do. In the end, things came to flashpoint, and David lost his temper and began to fight. In true Tom Sawyer tradition the two boys afterward became good friends. The boy had not fought particularly hard, David realized, and had simply wanted to befriend him. Custom demanded that such a friendship should begin in a physical way. The boy may even have been offended by David's initial refusal to tussle with him.

Despite their exterior roughness, David grew to prefer the Polish and Irish families to the members of his own Jewish household. He felt a great warmth when he visited their homes in contrast to the coldness and constraint of his own. They were willing to include him in their activities, sometimes inviting him to come to church with them or to eat a meal together. While he felt compelled to refuse, he was now feeling less of his own Jewishness. If he identified Jewish lore and customs with his father, then this was a way he would distance himself from Samuel. By the time he reached his late teens, he had become firmly agnostic.

In David's twelfth year, around 1929, the miners of Wilkes-Barre were hit by two disasters. The first was the spreading Depression; the second was an influx of cheap heating oil into New York City. The metropolis had always used hard coal from the Pennsylvania coal mines for its domestic heating, but New Yorkers were now converting to oil-fired furnaces. Although a few mines managed to stay open, the town was hit by a sense of helplessness. David now saw poverty in his friends' homes and sensed their shock that such a thing could ever have happened in America.

A few years earlier, he had been inspired by the great American Dream as he gazed at the setting sun on the U.S. half dollar.[14] It seemed to him that this dream had something to do with the power of the new, with the rejection of the old life of Europe, and that the farther west one traveled, the brighter the dream would burn. But now that dream had soured, and David had to think again about how social justice could ever be established.

For the rest of his life, he would remember the plight of the Wilkes-Barre miners. Their homes had been a point of stability in his life, but now even that had shattered. How easily irrational forces could sweep across society and break apart its order. If, in later life, David took the world's troubles onto his shoulders and dreamed of a life that could be ordered, harmonious, and free, the origins in part lay in his early desire to fight the forces that had destroyed so much that he loved in his childhood world.

Another element in David's general attitude involved the history of the Jewish people. Although he no longer accepted the Jewish religion, he retained a deep sense of the long history and fate of the Jews, a sense that he would keep for the rest of his life. The pogroms of central Europe and later the Holocaust, in which so many of his relations died, gave him an enduring fatalism that may well have led to his often-pessimistic vision of the human race.

Wilkes-Barre never recovered its earlier prosperity during David's childhood. As he progressed through high school, he took an increasing interest in politics and in the rise and fall of ancient civilizations. He invented interplanetary societies with histories of their own, projecting changes in their financial and other institutions for thousands of years into the future. He wondered if Western civilization could survive after the year 2000.

Still, his favorite topics remained mathematics and science, which he was able to study in greater depth during his final years of high school. One teacher in particular, Mayer Tope, brought mathematics alive for him. As he lectured, Tope would wave his Phi Beta Kappa key around, an idiosyncrasy that became something of a joke among the students. (Later, David would adopt the habit of swinging a key on the end of a chain as he talked.) At the age of ninety, after teaching for well over sixty years, Mayer Tope immediately remembered David Bohm. "Of the thousands of students I have seen in my life, David Bohm was outstanding. He made a tremendous impression on me. He was a very brilliant kind of fellow." Tope once set the class to solve a particularly difficult geometry problem, without much hope of anyone being able to solve it. To his surprise, not only did David come up with the solution, he did so in a highly original way.[15]

Sam Savitt, a fellow student, remembered that David solved the problem in three different ways, one of which he had to explain to Tope after school.[16] As Tope himself put it, "You can teach something in mathematics for many years, toiling away at it until you believe that there is no other way of doing it." But then a student like David Bohm would come along and show that it was possible to arrive at a solution by another route. "That was the sort of thing he did. Doing the impossible attracted him." On some occasions Tope would allow David to take over the class and cover the blackboards with mathematical equations. Although David was an outstanding student, Tope found him modest and not at all pushy—in short, a very good student to have in the class.

As David's interest in science developed, he became involved in deep discussions with one Rabbi Davidson. The rabbi was an educated and cultured man whose humanistic spirit complemented his religious nature. When David confessed to him that his overwhelming interest was in science and that he could no longer feel any connection to Jewish religion and its traditions, Rabbi Davidson asked him what he would have done had he been born in the Middle Ages. David replied that he would probably have been a very religious Jew, for one's values are set by the times and the society in which one lives.[17]

While Mort Weiss remained his close friend, David discovered other boys with whom he could feel comfortable. With them he would build model sailing ships and aircraft that never quite seemed to fly. Henry Kunicki was a close study companion for a time, until he was forced to leave school to work at the mines. Together Henry and David would discuss the harmony of the planets and the idea that all of nature could be subsumed under universal laws. If only human beings could realize that the entire universe was rational, David thought, then they would begin to apply reason and seek harmony in their own lives.

Meanwhile, back at the furniture store, Samuel Bohm was still insisting that his son get involved in outdoor activities. Most boys spent the summer at the local Boy Scout camp, but to Samuel's chagrin it did not provide kosher food. He bought a sixteen-foot tent so that David, Mort Weiss, Sam Savitt, and the rabbi's two sons, as well as Dai (David) and Kid (Jimmy) Jones, the sons of a Welsh miner, could camp out on their own for the summer. They set up camp at Harrison Park, in the Pocono Mountains, and periodically one of their relations drove up with kosher food.[18] At camp David exhibited a craving for sweet things, pies, cakes—anything made with sugar. The boys nicknamed him "sugar freak" because he spooned so much of it into his coffee. Finally they had to hide the sugar from him.

While other boys took comic books and sports magazines to camp with them, David arrived with a selection of textbooks. Mort had brought a light, attached to a large storage battery, so that he could read in bed. David immediately commandeered the device so that he could study throughout the night. Within a couple of days the battery was dead, and he had to wait to recharge it the next time Mort's uncle drove up. During those camp summers David always seemed to be reading so that he could pass extra examinations and take more advanced courses. When the other boys talked about sports, told dirty jokes, and speculated about sex, David preferred to talk about the future or work out a formula in his head.

Nevertheless, David enjoyed the outdoor life, and once he was away from home, he felt much healthier and stronger than he had imagined possible. If he could not play games, he could at least walk his friends off their feet, sometimes covering ten or twelve miles a day. On one occasion the boys had the use of an old Ford that they had to push in order to start. At that point they would run alongside and climb in the windows, yelling, "Let's go, Davie!" But David plodded along at his own pace, catching up with the car the next time it stalled.[19]

Back at school, David was coming to grips with more formal mathematics. In grade eleven, Mayer Tope taught the traditional Euclidian geometry of triangles, circles, and squares. What fascinated David was the idea of proof. Using pure logic, it was possible to prove that the angles of a triangle add up to 180 degrees. It was true not only of the one particular triangle on the blackboard but of any triangle on a flat surface, no matter how it was drawn. Proof seemed to extend to, and to anticipate, all possible worlds.

David was thrilled that the human mind could create logical patterns that were purely abstract yet would nevertheless apply to the real world of matter, space, and time. This was his first step toward formal science, that power of the mind to differentiate between what is logically necessary about the universe and what is merely conditional or contingent. It is the route by which universal laws are discovered, the relationships that underlie individual events and give order to the universe. Until he had that revelation, everything in David's world had been contingent and replaceable, but now he had discovered something that would stand fast.

David also studied algebra and was intrigued that a mysterious x could stand for anything at all. The human imagination could not only conceive of the unknown, he realized, but through reason alone, discover something about that unknown.

In the following year of school, his study of the Euclidian geometry of triangles, circles, and squares gave way to solid geometry in three dimensions—the properties of cubes, spheres, pyramids, cones, and cylinders. At last David had acquired the mathematical tools that would enable him to explore the idea of a fourth dimension, inspired by the magazine story. What if we are simply the three-dimensional cross section of some hidden four-dimensional world? he had asked as a child. Now he was in a position to show that what appear to be separate objects in our world could be, in four dimensions, aspects of a single whole. His ambition was, as he wrote in his notebook, to show that "all the apparent laws of the universe" have the same source. "This correlation has long been sought, and if it is true, I believe it will aid the future progress of science greatly."[20] In effect, he sought a cosmology, a scientific account of the interconnection of all things. In particular, David proposed to include the nature of mind within his cosmology.

Our universe of space, time, and matter, David argued, is the manifestation of an underlying four-dimensional "cosmos." This hidden world consists of an ether, which itself is composed of a vast number of infinitely small particles that are in constant agitation. Time is created by the uneven expansion of this ether, and it will run at different rates in different locations. Matter and gravity are also related to this ether in a way that is consistent with general relativity.

Although he had generalized the theorems of solid geometry to four dimensions, most of David's "general theory of the cosmos" was argued verbally and intuitively. It was an impressive thing for a schoolboy to have achieved, and it foreshadowed David's later work in its boldness of imagination, in its emphasis upon wholeness and interconnection, and in the notion that our everyday world is the manifestation of something hidden.

In his senior year of high school, David read about Niels Bohr's theory of the atom. This was the earlier form of quantum theory, in which electrons orbit around the atomic nucleus like planets around the sun. Not all orbits are possible, and to explain why, Bohr had proposed the existence of discrete, quantized orbits. David came up with his own solution: that electrons have tides, and stable orbits are formed only when the interval between successive tides is equal to the time the electron takes to circle around the nucleus. His ideas about electrons connected in his mind with vortices and tornadoes, that earlier preoccupation, and with the notion that objects could emerge out of pure motion.

During David's last years of high school, the news from Europe was increasingly disturbing. The rise of fascism worried him, particularly when he saw that it was being accepted by many Americans. In the local library he read copies of *Social Justice,* a magazine put out by Reverend Charles E. Coughlin of Red Oak, Michigan. In his Sunday broadcasts from "the Shrine of the Little Flower," Father Coughlin attacked the New Deal, "Jewish bankers," and Communists. When, in 1934, he launched his National Union for Social Justice, more than five million radio listeners signed up in two months. Father Coughlin was said to be more popular than the president.[21] Coughlin symbolized for David the rise of the right in the United States.[22] One reader's letters to *Social Justice*, he noted, said that it was not sufficient for believers to walk behind Christ—they should march ahead of him with a club!

Samuel Bohm and other local Jewish businessmen were blind to these dangers, it seemed to David. They even approved of Mussolini's new Italy. They took Hitler's anti-Semitic speeches as mere political rhetoric, joking that "the soup is never drunk as hot as it's cooked."

David tried to express his worries in a mock political speech he gave at school in support of a mythic fascist dictator, Adolph Staliney. The dictator's motto was "Staliney never Stalls," and his salute involved throwing up one's hands and emptying one's pockets of valuables. David hoped to use satire to make serious political points, but when he began to speak he was so stiff and tense that he felt the whole thing was a failure.

In reaction to the right, David began to read left-wing journals like *The Nation* and *The New Republic*. He was not yet convinced that socialism was the answer, but as he read, he realized that the adults he had talked to had misrepresented the left. He discovered the perfect foil for his ideas in Mort's father, Samuel Weiss. Samuel's uncle-in-law was A. A. (Abe) Heller, a friend of the industrialist Armand Hammer. Both men knew Lenin and were particularly sympathetic toward the Communist Party.*

Mrs. Weiss considered Uncle Abe a paragon, which so infuriated Samuel Weiss that he took a diametrically opposed position and had begun to play the stock market. Whenever David visited Mort, Mr. Weiss would extol the virtues of capitalism and claim that people were only motivated by greed, self-interest, and fear. David argued back, quoting from articles in the left-wing magazines. The two would still be going at it long after Mort went to bed, until around two in the morning, when David walked home. Occasionally a neighbor would drive past and offer him a lift, but David preferred to walk alone in the dark. He had a great deal of thinking to do. But with his time in Wilkes-Barre drawing to a close, two passions had been confirmed in him: physics and politics.

* As a young man, Hammer made his first million in pharmaceuticals, then, in 1921, traveled to Russia, intending to give medical aid to famine victims. He became friendly with Lenin, who explained to him the economic difficulties faced by the Bolsheviks and suggested that Hammer's business talents would ultimately be of more use than medical aid. As a result, Hammer established a factory in Russia to produce cheap pencils. Abe Heller, Hammer's friend, returned to the United States in 1930, subsequently making a fortune in petroleum. His interest in the Soviet Union continued, and during the 1970s he encouraged closer business ties between the United States and the USSR.

CHAPTER 2

Youth: From Penn State to Caltech

SAMUEL BOHM may not have thought much of "scientism," but at least he was willing to pay his son's way through college. David's choice was Pennsylvania State College, a relatively small school located in rural College Park, some 120 miles from home. While it was not particularly distinguished, David preferred to avoid academic competition, and it had been the college his maternal uncle Charles had chosen. As it turned out, he was far happier at Penn State than he would be at the more prestigious Caltech, which he later attended.

During the early decades of this century, physics was not taken seriously in the United States, whereas European universities boasted significant departments of physics. In Munich, for example, Arnold Sommerfeld lectured to a class that included Wolfgang Pauli, Werner Heisenberg, Herbert Fröchlich, Fritz Bopp, and Rudolph Peierls, all of whom were to distinguish themselves. At Penn State, however, the focus was on practical subjects like agriculture, engineering, and chemistry, and the small physics department justified its existence by teaching service courses to students from other departments. When Bohm arrived, he discovered that only a handful of students were serious about science, and the professors were engaged in little original research. Nevertheless, he rapidly developed his own course of independent study, which included wide reading and discussions with

Mort Weiss, who was studying electrical engineering, and with two or three of the better students.

During their first semester, Bohm and Weiss shared a room, where David would talk long into the night about space travel and alien intelligences until Weiss pretended that he had fallen asleep. It was during this period that David began to experience the stomach disorders that plagued him throughout much of his life. Initially, the problem was to obtain kosher food on campus. While David had moved away from the Jewish religion, he still kept to its dietary laws. At first he attempted to circumvent the prohibitions by refraining from meat. After a month of this regime, however, he felt unwell. His various stomach complaints were compounded by the fact that he ate in cheap restaurants such as Boots Diner, his favorite haunt. Finally, his addiction to sugar had not diminished. Weiss worked for a time as a cleaner at Boots and would bring home unsold fruit pies for David. On other occasions he would watch in wonder as his friend worked his way through a family-size tub of ice cream.[1]

Despite his atrocious eating habits and his uneasy stomach, Bohm was now feeling fairly healthy. Probably the most important change in his life had been to get away from his oppressive home, and while other students often hitched rides home on weekends, David rarely returned to Wilkes-Barre. He enjoyed the countryside around the college, and each day after lunch, he would take a two-hour walk, or a cycle ride, passing outlying farms and climbing into the mountains. While he was walking, he would think about physics, for only in the open air did he feel his brain was flooded with the oxygen needed for creativity. Bohm was particularly attracted to woods and admired their complexity and the dappled light that filtered through the branches. Trees were "wild and unconstrained" to him, yet they betrayed a subtle underlying order that he could see as he looked from trunk to branch and from branch to twig. Human civilization would have developed in a better way, he speculated, if it had not exchanged the forest for the "oppressive and depressing" city. When he returned from these hikes, his friends found him revived and cheerful.

During his walks Bohm pressed ahead with his idea of a four-dimensional cosmology, which he now believed was more imaginative than Einstein's general relativity theory. Although he did not possess the necessary mathematical tools to read Einstein's scientific papers, he had a general idea about the theory and viewed Einstein as a heroic figure, the exemplar of all that was best in physics.

Following his evening meal and a period of study, Bohm would take a shorter evening stroll with Weiss across the campus. One evening Weiss protested that it was cold outside and snowing hard, but Bohm insisted that they should go out nevertheless. As they trudged along, Weiss made snowballs and invited his friend to a fight. When Bohm objected, Weiss started to lob them through some open windows in a darkened dormitory. He expected lights to switch on and angry students to shout, but nothing happened. Inside, however, a group of students were gathering in a corridor. Weiss kept throwing snowballs until Bohm protested that the whole thing was silly, walked up the steps, and into the dormitory. At the door he was met by a group of young men, one of whom punched him in the face, breaking his nose.[2]

Many years later, when Weiss met his friend, he noticed the distinct bump on Bohm's nose and apologized for what he had done. He had been worried, he said, that the nose-break would restrict the famous "floods of oxygen," to the detriment of Bohm's creativity. Bohm laughed, and as the two men relived the event, Weiss recalled that his friend had never borne him a grudge.

Bohm's college years passed in relative calm and tranquillity. His energies were so focused on science that he did not bother to join a fraternity, date girls, or go to social events. He was, however, obliged to join the college's reserve officer training corps. As might be expected, Bohm was remarkably uncoordinated at drill, to the point where an officer told him, somewhat prophetically, "there are some people who just can't march with the others. You're one that will have to go on his own." By contrast, target practice proved enjoyable, and he developed skill at hitting the bull's-eye.[3]

In his third year Bohm teamed up with another physics student, Maynard Dawson, to work their way through the problems in Whittaker and Watson, a textbook that presented an overview of all the mathematical skills needed for research in theoretical physics, at least in the early decades of this century.[4] By solving these problems, Bohm was developing skills associated more with someone working toward a Ph.D. than those of an undergraduate. As it turned out, when he arrived at Caltech to begin graduate work, he found that his self-directed learning had given him far better foundations than students from larger and more prestigious schools.

By now, Bohm had begun to impress his fellow students. Harold C. Sebring, a chemistry major, was astonished at the speed with which Bohm could complete a calculus assignment, while Howard Greenwall, also majoring in chemistry, was glad for the advice Bohm offered him in his own subject. Bohm's scientific precocity was also apparent to his professors. During a third-year course on advanced heat, a proof was given that radiation exerts a pressure. In the following lecture the professor announced that "Mr. Bohm" had explained to him that the proof, which could be found in a number of textbooks, was in fact based on circular reasoning. "Mr. Bohm" had talked it over with him, and they had developed a new proof, which Bohm agreed was correct.[5]

On another occasion Bohm discovered an error in an electronics textbook, but when he pointed it out, he was reprimanded by his professor, who told him that "it's not the place of students to point out errors." On the following day, however, the professor had the good grace to apologize.[6]

Moving among students with a wider range of backgrounds and persuasions than he had previously encountered, Bohm's interest in politics flowered and, at times over the next years, would "rise to passion."[7] In high school he had been interested in the sweep of history and the rise and fall of civilizations; he had argued politics with Mort Weiss's father and the local rabbi. Nevertheless he had still retained his faith in the power of democracy and Roosevelt's New Deal. But now, as he took stock of the political groups around him, from the Young Communist League on the left to the Nazi-saluting German-American Bund and Father Coughlin on the right, he became more cynical. It seemed to him that people only paid lip service to the ideals of democracy and freedom but were privately motivated by self-interest, while powerful groups were willing to employ repressive measures to attain their ends. In short, America seemed fertile ground for the rise of fascism.

That conclusion made Bohm pay more attention to what the

radical students were writing in the campus newspaper. Some form of socialism seemed to be the answer; after all, during the Depression it had been necessary for the state to intervene in economic life and prime the economic pump. Yet he was not yet totally convinced, for at times the left appeared overly dogmatic and leaned toward Communism and the Soviet Union. A new society was required, Bohm realized, but it had to come about in a rational way, which for him meant by the use of science. The scientific mind was the proper response to the universe, and human beings would be able to understand themselves only when they understood the universe around them. Science offered the possibility of such understanding, as well as the ability to control nature, using its forces in the service of the human race. The political system Bohm sought must be strongly based in science, and its goal was an orderly, rational society in which human freedom was guaranteed.

In his senior year at Penn State, Bohm devoted less time to formal coursework, preferring to spend most of his days walking, thinking, and reading. With his future to consider, he applied to graduate schools that would offer assistantships to pay his fees and expenses. Although his grades had been good, the replies he received were far from encouraging. Only later did a family member tell him that, while the head of the physics department had given him a strong reference, he had also taken care to point out that Bohm was a Jew![8]

Another possible funding source was Penn State's mathematics prize. The contest consisted of five questions that had to be answered in one afternoon. Most of the students managed to get through one and a half questions, and anyone who successfully completed two would generally be the winner. During that afternoon Bohm finished four of the problems and drafted an outline for solving the fifth. That year a story circulated on campus about the way the scholarship was awarded. A certain mathematics professor who was notorious for his total lack of involvement in all university affairs attended the scholarship meeting simply to express his support of Bohm.[9]

Finally two letters of acceptance arrived. One was from the State University of New York at Rochester. Bohm joked that since the head of physics there was a Jew, he was presumably not prejudiced against having other Jews in his department! The other was from the California Institute of Technology, which would have been Bohm's choice except that no assistantship was offered. He had always dreamed of traveling toward the setting sun to California, the point farthest from European influence. Caltech was particularly inviting for him since Einstein and other noted physicists had spent time there. Yet Bohm would be unable to go without financial help. One night as he was pondering his future, he experienced such an extraordinary surge of energy that he was unable to sleep. At sunrise he set out on a very long walk, and on his return, he found that a letter had arrived from Penn State awarding him a fellowship of six hundred dollars, which he could spend anywhere he liked. His decision was immediate: he would go to California.

His degree completed, Bohm returned to Wilkes-Barre to spend the summer of 1939 at home. As for his trip west, he replied to an advertisement offering to sell the return portion of a bus ticket to California for only thirty dollars. The day, in September 1939, when Bohm boarded the bus was particularly significant: War had been declared in Europe. Arriving in Los Angeles, he took a room at the local YMCA. There he met Leon Katz, a physics student from Ontario. They got on well together, and since Katz knew of a good room in Pasadena at a moderate rent, he and Bohm decided to share it.

Bohm had finally reached his California dream, yet his first impressions were far from favorable. Pennsylvania had meant the shaded, dappled light of woods and forests, but here he was overwhelmed by a desert of unremitting light and heat. Only later, when he climbed into the mountains, was he able to walk within the shade of trees. And if the countryside was a bitter disappointment, the college was not much better. Bohm had hoped to continue as he had done at Penn State, reading textbooks, discussing physics with interested students, having contact with helpful professors, and beginning research work of his own. Instead he faced a heavy load of coursework, frequent examinations, and an all-pervasive spirit of competition. As far as he could see, his professors were more interested in teaching students how to solve problems than in explaining the underlying principles of physics.

Bohm could see no creativity, no intellectual freedom in these courses, and he became so depressed that his negative feelings colored everything around him. Even his fellow students, he felt, had been driven to conformity by the pressures of excessive competition. All that science seemed to mean to them was a way to make money and get ahead. Their attitudes were cynical, and they had little interest in the larger political movements sweeping Europe and the United States. Bohm was particularly irritated by the student elections, which had become mere occasions for absurd speeches, blue movies, and the distribution of cigars and condoms. While pressures of coursework were partly to blame for this frivolity, the attitudes of his fellow students only reflected those of their parents, he believed—indeed, the larger political skepticism within the United States.

Bohm was always the idealist, someone whose far-ranging dreams and expectations could be—and sometimes were—bitterly dashed. Again and again throughout his life he would travel to a new location, or enter a new university, in a spirit of hopeful anticipation, only to swing back into disappointment and even depression, which would then cast its grayness over everything around him.

His roommate, Katz, for example, felt very differently about "the Caltech style." He pointed out that each week the professors expected the class to study a section of the set text, along with a series of problems. The following lecture would commence with a discussion of the previous week's problems, after which the professor was free to lecture on anything that took his fancy. Katz believed that Caltech students learned physics through the act of problem-solving itself. But for Bohm, understanding always involved probing deeper and deeper into underlying assumptions.[10]

Despite their differences in approach, the two students worked together on their assignments. According to Katz, Bohm did most of this work and was usually able to polish off the assignment in around an hour, spending the rest of the evening immersed in physics texts. Both of them took the legendary Electricity and Magnetism course taught by W. R. Smythe from his textbook *Static and Dynamic Electricity*.[11] On more than one occasion, Smythe would copy Bohm's answer onto the blackboard, remarking that he had never seen the problem solved in that particular way before. Bohm acquired the reputation of being the first student in Caltech's history to solve every problem posed by Smythe.

Bohm and Katz also attended the course given by Fritz Zwicky, the astrophysicist, who spoke, it was said, a number of languages with a very heavy accent, including his native Russian![12] At that time Zwicky was engaged in research on astronomical nebulae, and after assigning textbook problems to the class for several weeks, he suggested that they might like to work on a real problem. He outlined the conditions inside nebulae and pointed out how, in their random motions, the elementary particles within these clouds collide with each other. Random collisions between pairs of particles were frequent. But what was the chance, he asked, of a three-body encounter—that is, of a third particle coming along at the exact instant when two other particles collide?

Bohm and Katz struggled with the problem, and by the end of the week, they were the only ones to have gotten anywhere. After setting down their work on the blackboard, Zwicky encouraged them: "Good, good. Well, continue with it." At each lecture Bohm and Katz reported on their progress until one day Bohm wrote the final solution on the blackboard. At this Zwicky looked into the notebook that he always carried and said, "Yes, that is correct. Now I can publish my paper."[13]

Despite these displays of brilliance, Bohm realized by the end of the first semester that he could never fit into the Caltech atmosphere or benefit from what the institute offered. The constant problem-solving depressed him, and the pressures of quarterly examinations seemed oppressive. Possibly he may even have suspected that he was out of step with the mainstream of physics and was destined to become a maverick, a loner, at odds with the current trends in physics. As such, Bohm was really a throwback to an earlier age, in which physics involved deep and quiet contemplation of nature; when it was more concerned with discovering the underlying order of cosmos than with making predictions and solving practical problems. When he thought back to the golden age of Planck and Einstein, or even earlier, to the age of Newton, Bohm could not help feeling that physics had become small and the concerns of its practitioners petty.

To a physicist of the caliber of Richard Feynman, an area of physics was interesting only if he could find a problem in it, and he raised his art of problem-solving almost to the level of genius. While occasions arose in his own research when Bohm was forced to solve technical problems, he always distrusted abstract mathematical proofs. After all, he would say, there are always unexamined assumptions in any piece of mathematics, and the more complicated the mathematics, the easier it is for undetected errors to creep in. Rather than proceeding in a relatively mechanical or logical way, he preferred to feel out the answer and see it in his mind before setting down the necessary mathematical steps. It was as if his problem-solving ability were guided less by logic than by a combination of imagination and intuition.

While he was at Penn State, for example, he had been trying to understand the theory of the gyroscope, the toy that intrigues children because of its ability to remain in balance. Normally when an object is pushed, so that its center of gravity moves outside the point of balance, it falls. A gyroscope, however, does not fall. Instead, its axis of rotation moves—that is, it precesses. Try to push a gyroscope in one direction, and it will react by moving in a direction at right angles.

Faced with explaining gyroscopic motion, most physics students learn the various formulae, involving conservation of angular momentum, and produce an explanation in a relatively mechanical and formulaic fashion; but Bohm needed a direct perception of the inner nature of this motion. Once as he was walking in the country, he imagined himself as a gyroscope, and through some form of muscular interiorization, he was able to understand the nature of its motion. In this way he worked out, within his own body, the behavior of gyroscopes. The formulae and the mathematics would come later, as a formal way of explaining his insight.[14]

From very early on in his scientific career, Bohm trusted this interior, intuitive display as a more reliable way of arriving at solutions. Later, when he met and talked with Einstein, he learned that he too experienced subtle, internal muscular sensations that appeared to lie much deeper than ordinary rational and discursive thought.

Without explicitly knowing it at the time, Bohm had returned to that ancient maxim "as above, so below," the medieval teaching that each individual is the microcosm of the macrocosm. Bohm himself strongly believed himself part of the universe and that, by giving attention to his own feelings and sensations, he should be able to arrive at a deeper understanding of the nature of the universe.

This particular skill remained with Bohm throughout his professional life. His colleague at Birkbeck College, Basil Hiley, once remarked, "Dave always arrives at the right conclusions, but his mathematics is terrible. I take it home and find all sorts of errors and then have to spend the night trying to develop the correct proof. But in the end, the result is always exactly the same as the one Dave saw directly."[15]

Toward the end of his first year at Caltech, then, the only pleasure Bohm obtained from physics was from his own private reading. He spent several weeks studying Paul Dirac's elegant and formal account of the new quantum theory. He had also come across Arthur Eddington's *Relativity Theory of Protons and Electrons,* which became his constant companion for several months. Eddington, a noted British astronomer, mathematician, and physicist, had led the 1919 solar eclipse expedition to West Africa, where measurements had confirmed Einstein's prediction that starlight is bent by the sun's gravitational field. In effect, Eddington had established the experimental credibility of the general theory of relativity. In addition, he wrote an excellent exposition of Einstein's theory.

In 1936 Eddington's *Relativity Theory of Protons and Electrons* offered a unified theory that linked relativity to the world of elementary particles. It was inevitable that Bohm was attracted to this all-embracing theory of the universe. Here was a thinker whose deep understanding of physics, imaginative flair, and overarching dream of a totally unified physics spoke directly to Bohm. The only problem was that the book was so highly mathematical that Bohm had to struggle to understand the theory. To recast it in his own terms, he wrote out a nonmathematical version of the theory, which he showed to one of his professors. The professor dismissed it, saying that no one really understood Eddington, and anyway it was not at all clear that there was anything to his theory.

Bohm felt that he had been a fool to get excited about nothing, and with the help of a friend, he wrote a satirical story about Arthur Eddington. It took as its starting point Eddington's dramatic claim that he could deduce the nature of reality unambiguously from epistemological considerations alone. This caused such concern in Hell that the Devil paid a visit to Eddington and made an offer for the scientist's soul. The Devil was generous, proposing money, power, and women.

Eddington refused, which came as a great surprise to the Devil, for "most scientists will sell their soul for something smaller, like a navy contract. We're offering you the whole works and still you won't sell. Can you at least explain yourself?"

Eddington replied that he had deduced the nature of the universe unambiguously on epistemological grounds alone—if he wanted anything, he could have it.

At this the Devil admitted that he did not understand Eddington's book. "I have had my best scientists working on it, and they can't understand it. It is very urgent that I should understand the nature of reality in God's world." He asked Eddington to explain the meaning of his book. So insistent was the Devil that he finally offered the scientist his own soul. Eddington agreed. And so, Bohm concluded, if you want to understand Eddington's theory, you can go to the Devil![16]

Because he was spending more and more time away from his regular assignments, Bohm's marks had suffered (albeit only in subjects that did not particularly interest him), and he was told that he would not be offered an assistantship for the following year. As Katz was preparing to leave to Canada for summer vacation, he noticed how depressed his roommate had become, and that his stomach problems had flared up again. "If Dave didn't go to the toilet at the right time," he later remarked, "it would upset his whole day."[17]

With Katz in Canada, Bohm took a room in a house with a Chinese student. Chinese culture represented one aspect of Bohm's California dream, and for a time he studied its language and philosophy enthusiastically. As he talked and read about the detachment and the philosophy of inaction developed by the ancient Chinese, he wondered if something similar was developing in the United States, as the energy of its European heritage burned itself out.

Following a short visit home, Bohm was back at Caltech for the new school year. But now his despondency could be relieved only by long hikes. His favorite climb was up Mount Wilson, which rises to 5,710 feet; and the path Bohm took from the campus to its summit rose 3,000 feet. It was certainly not the sort of climb that a sickly person could undertake on a regular basis. Until the last years of his life, Bohm complained about his health, but his energy for walking—and for talking about physics—was boundless.

During this second year Bohm had to do his first serious research project. The topic given to him was to calculate the way interstellar dust clouds and nebulae scatter light. It did not appear particularly exciting, and neither did Bohm get on well with his supervisor. The problem was straightforward, the sort of thing that a graduate student can solve in a relatively mechanical way. This approach, however, did not satisfy Bohm, who racked his brain for alternative and more imaginatively satisfying approaches. The result was that he spent most of his time thinking about the problem instead of getting down to the business of solving it. After several months of wasted time, he was forced to admit defeat and slogged away at the details of calculation. If this was physics, he felt, he would have been far better off working in a furniture store, as his father had wanted.

Clearly the situation could not continue. It was around this time that a friend suggested Bohm should approach J. Robert Oppenheimer. Several years earlier, Oppenheimer had established a school of theoretical physics at the Berkeley campus of the University of California, which was expected to become the crown jewel of American physics. Bohm met with Oppenheimer, who seems to have been sufficiently struck by the young physicist that he promised to arrange something for him at Berkeley.

With great relief Bohm dropped out of Caltech. His records at the transcript office show his date of entry into the institute (1939) but no official date of departure. Bohm moved from Pasadena to Berkeley, staying at the YMCA while Oppenheimer made the arrangements necessary for him to enroll at the University of California. Already Bohm's spirits were lifting, the climate was more temperate, and he could walk in forests of giant redwoods. In 1941, with the help of a small assistantship that only just covered his living expenses, he was working in Oppenheimer's theoretical group. The years that followed represented a flowering of his creativity. At last he had found the scientific home of which he had long dreamed.

CHAPTER 3

A Vision of Light

BEFORE THEIR FIRST MEETING in the early 1940s, Bohm had already been well aware of J. Robert Oppenheimer, the young American genius whose physics department rivaled anything in Europe. Bohm and Oppenheimer shared Jewish origins, and as children David and Robert had had significant characteristics in common; both were melancholic and introspective, their lives were dedicated to books, and they disliked organized games. But there the resemblance ended, for nothing could have been further from the Bohms' furniture shop in Wilkes-Barre than the Oppenheimers' Riverside Drive apartment in Manhattan, hung with paintings by Van Gogh, Cézanne, Derain, and Vuillard.

Oppenheimer was born on April 22, 1904. Ella, his mother, had a sensitive nature, and after studying art in Paris, she had worked in New York as a painter and art teacher. Following Robert's birth, she was forced to give up her profession, pressured to conform to the mores of the society in which her husband, Julius, moved. Julius, for his part, was ambitious and had strong social aspirations. Deciding that his son's name, Robert Oppenheimer, was insufficiently distinguished, he added the initial of his own first name, thus transforming his son into J. Robert Oppenheimer.[1]

The atmosphere at home was claustrophobic, formal, and strained, run by servants; young Robert dressed each evening for dinner. He was driven to school by a chauffeur, where he insisted on waiting for the elevator in order to be conveyed to the upper floors, thereby holding up classes. The boy showed a precocious interest in science and, at the age of twelve, gave a paper to the New York Mineralogical Club. He wrote dreamy philosophical poems and took a great interest in literature. Achieving the highest grades in his school, he developed an intellectual arrogance and a snobbish aloofness that remained with him throughout his life, making it difficult for him to find friends. He embraced a strict code of manners and was appalled when those around him violated it.

At Harvard University the young Oppenheimer excelled at Greek, French literature, and chemistry, so much so that it was not at all clear where his talents should be directed. Only when he attended the classes of Percy W. Bridgman, an inspiring teacher as well as an outstanding philosopher and experimental physicist, did Oppenheimer become fired with a passion for physics. From that point his path in life was clear, and after graduation, he left for Cambridge, England, with a letter of introduction to Ernest Rutherford in his pocket. The outstanding scientist of his day, Rutherford had discovered the structure of the atom, and his Cavendish Laboratory, at Cambridge University, was the mecca for aspiring physicists. Yet once settled in England, Robert became homesick and bored with experimental research. Frustration gave way to a depression so severe that he contemplated suicide. Within a few months his mental instability was so extreme that, while visiting Francis Fergusson in Paris, he fell upon his friend and tried to strangle him.

Robert's condition was diagnosed as dementia praecox (the old term for schizophrenia). The cure for his condition he found not in the ministrations of the medical profession but in a vacation in sunny Corsica, combined with a passionate but secret love affair. Back at Cambridge, his depression lifted, he met Niels Bohr, and he decided that he would become a theoretical physicist.

The period was a revolutionary one for physics. In 1925 Werner Heisenberg had discovered quantum mechanics, and within a few months Erwin Schrödinger had developed wave mechanics, based upon Louis de Broglie's ideas of matter waves. In 1926 Oppenheimer was in Göttingen, studying with Max Born and meeting the leading players in the new quantum theory. Among visiting scientists of the caliber of Heisenberg, Wolfgang Pauli, and Paul Dirac, Oppenheimer was regarded as a show-off and an intellectual snob. Not content with working at the forefront of a scientific revolution, he was also studying French literature, talking knowledgeably about music, and learning Italian in order to read Dante and write Italian poetry of his own. To Dirac, this diffusion of attention merely indicated a lack of purpose.

It was almost as if Oppenheimer could grasp the essentials of any subject without having to spend time studying its details. Later in life, when he had students and a department of his own, his surface brilliance burned as bright as ever. He once told his assistant, Melba Phillips, that but for physics, he would have been a dilettante. David Bohm went further: "He *was* a dilettante. He just would not take his coat off and really get stuck in. He'd got the ability certainly. But he hadn't got the staying power." Another of his students, Joe Weinberg, compared Oppenheimer's mind to a mountain goat that makes sudden leaps that others are unable to follow.[2]

Scientists in the United States urged this brilliant young man to return to his native country and create a center of excellence there. He accepted a position at the University of California, Berkeley, which, thanks to its staggered terms, would also enable him to teach at Caltech each spring. But before he took up these positions in 1929, he returned to Europe, this time to work with Wolfgang Pauli in Zurich.

Back in California, Oppenheimer gathered about him a circle of exceptional research students. His talk was of physics, wine, Eastern philosophy, Sanskrit texts, quantum mechanics, and French poetry. It was said at Berkeley that "Bohr is God and Opjie is his prophet." (*Robert* to his close friends, Oppenheimer preferred to use the Dutch nickname *Opjie* with his students. The form *Oppie* was used by those who knew him less well.) While his early lectures frustrated his students with their mumbled delivery and rapidly vaulting thoughts, by the time Bohm arrived, Oppenheimer had become a brilliant teacher. Joe Weinberg felt that he always directed his lectures to the very best in his audience, making interconnections, allusions, and suggestions while throwing out ideas in full flight. Students of lesser ability found these performances discouraging, but others were inspired by them.

Prior to Bohm's arrival—and Oppenheimer's marriage—the professor would talk to his students long into the evening, sometimes taking them to an expensive restaurant in San Francisco or drinking with them until it was time for the last ferry across the bay. His parties were famous and, according to ritual, ended with one of the late Beethoven string quartets being played on the gramophone. These compositions, in their complexity, reminded the physicist Isidor Rabi of Oppenheimer himself. Rabi used to joke that he had spent much of his life constructing theories about Oppenheimer, only to discover finally that they were incorrect.

With the help of a small assistantship, Bohm joined this group and was immediately struck by the contrast with Caltech. Not only were the Berkeley students superior, having a much wider range of interests, but the atmosphere was relaxed. Above all, Bohm was captivated by Oppenheimer. His first impressions were of an immensely exciting personality with such an intense interest in scientific and philosophical ideas that his students could not help but be swept away by him. The atmosphere he created was lively and vivacious, so bursting with ideas that some students found it difficult to keep their balance.

Bohm's feelings for Oppenheimer extended beyond admiration into what he later described as love. Here was someone who not only understood the passions of Bohm's intellect but who offered him encouragement and support. It was inevitable that a part of Bohm's nature would look to Oppenheimer, thirteen years his senior, as a protective and understanding father.

Bohm's first months at Berkeley had the flavor of one of those grand tours of the Continent undertaken by young Englishmen in the eighteenth and nineteenth centuries. A boy from repressed and narrow Wilkes-Barre was suddenly introduced to the cultural world outside physics. He was encouraged to explore philosophical, social, and political issues and to consider the wider implications of physical theories. How dull other universities and their staffs would appear after his years at Berkeley!

Even the physical environment was exhilarating. In Pasadena Bohm had been oppressed by the heat and light, but now he lived in a temperate climate surrounded by trees. He took a room in a large house close to Strawberry Canyon and agreed with the woman in the next room who renamed it Storybook Canyon, in memory of the magical tales of her childhood. Walking along that canyon, he would pass E. O. Lawrence's cyclotron, then climb Grizzly Peak to look out across San Francisco Bay. Later in life, he compared his own scientific position in the face of the cosmos to that of the graduate student who once stood on Grizzly Peak.

Isaac Newton had likened himself to a child picking up shells on the shore, while lay the vast uncharted ocean of truth nearby. Bohm's self-images returned to his childhood, to the sun setting behind mountains on the half dollar, or to his walks around Penn State as he dreamed of traveling west. He thought of the original Berkeley—George, bishop of Cloyne, the philosopher and educator who had written, "Westward the course of empire takes its way." Now he was at the New World's most westerly point. Normally reserved and inhibited, he experienced such freedom as he looked over the ocean that his feelings were unchecked. Standing alone on Grizzly Peak, he felt totally involved in nature and sensed the flowering of his creativity.

When it was time for Bohm to begin his research project, Oppenheimer suggested that he study what happens when protons and deuterons collide. The problem he was given had applications in everything from the inner structure of elementary particles to reactions inside stars and would also apply to the theory of the hydrogen bomb.

To undertake his research, Bohm had to draw upon the developments in physics during the first decades of the twentieth century. At the end of the nineteenth century, the English physicist Joseph John Thomson had pursued Sir William Crookes's investigations of electrical discharges in gases and confirmed that electrical currents are indeed produced by the movement of tiny negatively charged particles he called electrons. A few years later Wilhelm Wein and J. J. Thomson discovered a heavier, positively charged particle called the proton.

The obvious conclusion was that the atoms that make up all of matter are composed of protons and electrons. (The electrically neutral neutron was not known at that time.) The problem was how to look into an atom and determine its detailed structure. Tiny objects, such as living cells and fragments of rock, can be examined under a microscope, but light is far too gross a tool to probe such a tiny entity as an atom. The wavelength of light—the distance between one peak and another—is vastly greater than the dimension of an atom. Light simply does not "see" atoms.

If atoms are invisible, then how were physicists to deduce their inner structure? It was that genius of experimental physics, Ernest Rutherford, who provided the answer. In the 1920s he began to shoot charged nuclear particles into atoms. His reasoning was that when a positively charged "bullet" approaches a positively charged proton, the "bullet" will experience a strong electrical repulsion that will deflect it from its path. Measuring the degree of this deflection over a very large number of experiments allowed Rutherford to plot his map of the internal structure of the atom.

Rutherford had assumed that electrons and protons are dotted around an atom like fruit in an English Christmas pudding. He expected to see many small deflections in the paths of his nuclear "bullets." To his surprise, however, most of the "bullets" passed through the atom undeflected; only a very few were deflected, and those by a large degree. The inevitable conclusion was that, rather than being uniformly distributed all over the atom, the protons are all concentrated into a tiny central region—the atomic nucleus, as it was to be called. Most of his "bullets" had missed this nucleus, passing through the atom unscathed; only those few that came close to the tiny central nucleus experienced the cumulative effect of many positive charges, which scattered an incoming "bullet" violently, shooting it back toward the experimenter.

In this fashion Rutherford was able to map the interior structure of the atom. He also established the basic experimental technique for exploring subatomic nature, one that continues to be used, albeit in more highly sophisticated ways, at present. Physicists learn about the internal structures of elementary particles by shooting them at each other, then observing how they scatter. In order to interpret these experimental results, a comprehensive theory of scattering is needed. One aspect of this theory was the research problem given to Bohm.

In one important respect Rutherford's first model of the atomic nucleus was incomplete, for physicists subsequently discovered the electrically neutral particle called the neutron. Atomic nuclei contain roughly equal numbers of protons and neutrons. Bohm's problem involved the scattering of the simplest of these—the nucleus of the hydrogen atom, which contains a single proton—in collision with the deuteron, a tightly bound proton-neutron composite that forms the nucleus of the heavy hydrogen atom. (Heavy hydrogen, or deuterium, is a component of heavy water, whose formula is D_2O instead of the more familiar H_2O.)

As in all research work, Bohm's first step was to visit the library to determine whether anyone else had tackled the same problem. It turned out that the English physicist H.S.W. Massey had already tried, but Bohm could make little sense of his method. (A few years later Massey himself was to arrive at Berkeley as the leader of a British contingent of physicists.) Bohm was, however, happy to be working on a satisfying problem. According to habit, he did most of his thinking during his daily walks, and it was on his return from one of these, on December 7, 1941, that he learned that Pearl Harbor had been bombed by the Japanese. The news shocked him into the realization that over the past months he had been focusing, to the exclusion of everything else, on his research.

In the spring of the following year, Oppenheimer asked Bohm to report on his progress. Over the next two weeks he prepared intensively for his seminar. When the day arrived, he had worked himself into an extreme focus of concentration and, as he put it, had built up a great charge of mental energy. A naturally shy person, as he began to speak he nonetheless felt that everything was going extremely well. Soon he had the sensation that he was going beyond physics into something almost mystical, to the point where he felt himself in direct contact with everyone in the room. He was convinced that each individual consciousness had been transcended so that his audience was also sharing this experience. His impression was of an intense burning light. Again, as in childhood, light was the symbol of the world he had temporarily entered.

Did anyone else notice anything exceptional? His friends, Joe Weinberg and Rossi Lomanitz, would certainly have attended that seminar, but neither of them could recall that particular talk or indeed any seminar in which there had been unusual excitement, let alone a deep sense of connectedness.

Bohm described his experience in language more usually associated with mystical states. Yet the enlightenment experienced by a

Jakob Boehme or a William Blake resonated throughout their lives, informing the qualities of the world around them. In Bohm's case, he was rapidly plunged from "the world of light" into "the world of shadows." Over the days that followed, whenever anyone attempted to congratulate him on his seminar, he deprecated his work. The contrast between that hour of intense illumination and his daily routine overwhelmed him, for now everything appeared humdrum, dull, and meaningless.

There are languages other than that of mysticism to describe such experiences. A psychiatrist may talk in terms of a fragile ego being flooded with highly charged symbolic material that it is unable to metabolize. The result is a temporary inflation, a feeling of omnipotence, that is often followed by a reactive depression. In Bohm's case, the depression lasted for almost a year. During this period he was unable to continue with creative research and sought psychiatric help.[3] Oppenheimer became discouraged with him, probably taking Bohm's self-deprecation as a personal affront, a dismissal of the research project he had chosen for his student.

It may have been during this time that Bohm's view of Oppenheimer changed. Bohm had always felt that a few of the students adulated Oppenheimer excessively. It also struck Bohm that his supervisor enjoyed this attention to the point that he needed his students' approval. Nevertheless, Oppenheimer himself was sufficiently intelligent to understand the shallowness of this psychological need and to desire something more sincere. Bohm cynically suggested that Oppenheimer wanted to be worshiped in such a subtle and intelligent way that it would appear entirely genuine.

This man whom Bohm loved and admired had other character defects that ultimately contributed to his own destruction. What Oppenheimer himself termed his "beastliness" occasionally surfaced in his dealings with students. A new arrival would often become the object of considerable attention on Oppenheimer's part. It was as if that young person were being seduced into believing that he or she was exceptionally talented, one of the best students ever to have entered graduate school. Oppenheimer would continue lavishing such attention until, out of the blue, during a seminar perhaps, he would lash out with a cutting, critical remark. He was not simply attacking a scientific idea in an impersonal way, but focusing disdain upon a person with the sole object of belittling them. For the recipient, the experience was so disorienting, painful, and paralyzing that in some cases it destroyed the desire to continue in scientific life. Nor did Oppenheimer always confine such attacks to his students—when made in public life, they earned him powerful enemies.

What was the reason for his sudden outbursts? Some of his colleagues believed that it was the frustration of a mind always several steps ahead of its audience, so that in pure irritation he would hit out, rejecting a person's idea as stupid and their abilities as commonplace. Bohm, who was impressed by the speed with which Oppenheimer's mind could cut to the very center of a problem, understood how he could become irritated by the slowness of others. Although he was now seeing Oppenheimer more clearly, however, he continued to relate to him as a father figure. The final act of betrayal lay in the future.

Gradually Bohm's depression lifted, and he returned to his research. While he pushed ahead with the problem that Oppenheimer had assigned him, he also germinated an idea that was to resonate through his mind for the next few decades.

An outstanding feature of quantum theory is the idea of the dual nature of the electron, proton, or any other elementary particle: at times it behaves like a wave, at others like a particle. When billiard balls hit, a physicist can compute the exact paths taken by the balls before and after the collision. This cannot be done with elementary particles in quantum theory. Quantum theory speaks not of protons speeding away from a collision on definite paths, but of expanding waves. The density of such a wave in a certain region gives the probability of finding a proton in that region.

A scattered proton thus behaves like a spreading wave. Yet whenever we try to detect that proton, using a Geiger counter, it is always found in some definite region of space. Could it be, Bohm speculated, that the detection of the proton takes the form of a wave that collapses inward from all space? But if so, it would mean that the proton—or the electron, for that matter—is not a well-defined particle at all; rather, it is a process, a continuous process of inward collapse and outward expansion. Therefore every elementary particle collapses inward from the whole of space. In fact, each elementary particle is a manifestation of the whole universe.

Bohm's speculation may have had more than a little to do with his experience of his own consciousness, particularly on the day of his seminar. On that occasion he had felt his consciousness expanding, flowing outward into the room beyond its normal personal limits until it entered into a connection with others. Then as the seminar ended, it had collapsed back inward into depression.

Bohm's consciousness had expanded in other directions as well, for he was now taking a serious interest in women. One of his first girlfriends was a young student called Betty Goldstein, who described herself as being "fresh and naive." Later, as Betty Friedan, she would achieve fame with her book *The Feminine Mystique,* but at that time feminism was still, as she put it, a dirty word. When she knew Bohm, she was certainly not thinking about the liberation of women, and neither did she take physicists particularly seriously.

On arriving at Berkeley, Betty was amazed by the radical nature of the campus politics. She was sympathetic to it and joined a political study group with Bohm. For a time he was her boyfriend, and she remembered him as brilliant, shy, and attractive. On some nights she would look out of her window and see him in the street below, staring up at her room.[4]

During his years at Penn State and Caltech, friends had viewed Bohm as somewhat asexual. It now appeared that his problem had been inexperience and shyness more than lack of sexual drive. He was to have other girlfriends after Betty. One of them, named Helen, meant a great deal to him. They became particularly close and may even have contemplated marriage. For whatever reason, however, tensions surfaced and in the end they parted. This breakup caused such pain to Bohm that he talked about Helen for the rest of his life. A close friend during the 1950s, Miriam Yevick, speculated that he never fully recovered from the breakup of this early relationship, the trauma of which may have stunted Bohm's emotional development.[5]

Bohm was also building friendships with Oppenheimer's students, such as Richard Feynman, Rossi Lomanitz, Bernard Peters, and

Joe Weinberg. Bohm and Lomanitz roomed together for a time. Lomanitz found Bohm to be withdrawn even to the point of depression, yet once he could be drawn into a discussion of physics or politics, he would forget his bashfulness and become totally engaged.

Joe Weinberg was married and slightly older than Bohm. He had originally planned to be a writer and had edited a number of literary magazines; his other great love was music. Yet despite his literary and musical passions, the Depression years had convinced him that he needed something practical to fall back on, so he had enrolled in an engineering degree. While attending lectures on physics, he became excited by the field and decided to make his career in it. Weinberg was also passionate about philosophy, and by the time he arrived at Berkeley, he had developed an interest in Bohr's Copenhagen interpretation of quantum theory.[6]

Bohm often visited Weinberg's house to talk about physics, philosophy, and politics. There Weinberg, who owned a record player, introduced his friend to music. When Bohm first heard Beethoven's "Emperor" Concerto (Piano concerto no. 5), he had the sensation of "rising to the absolute top." The music also gave him an insight into Oppenheimer. Although his mentor was enormously talented, he had never made a truly outstanding advance in physics or exhibited that flash of genius that wins a Nobel prize. As he listened to Beethoven, Bohm realized that despite the power of his intellect, Oppenheimer lacked the ability to reach the summit.[7]

During the concerto's slow movement, Bohm experienced feelings he could only describe as religious. These feelings surprised him for he had by now rejected all taint of religion. Yet something in Beethoven and Mozart allowed him to touch the transcendent. Music made him realize the extent to which he had been leading a dulled and apathetic life.

The evenings he spent with Weinberg also brought him face to face with quantum theory. At Caltech Bohm had developed an antipathy toward quantum theory and was, as Weinberg discovered, "a convinced classicist." Classical physics, for Bohm, was clear and unambiguous, and even its most difficult concepts could be expressed in everyday language. By contrast, quantum theory seemed confusing and its interpretations, at times, arbitrary. Over the weeks that followed, like Arthurian knights, the two men did battle without offering quarter, each attempting to convince the other of the strength of his position.

Since Ernest Rutherford's discovery that atoms consist of negatively charged electrons orbiting around a tiny, dense, positively charged nucleus, physics had undergone a revolution. Picturing the atom as a miniature solar system posed a serious problem for classical physics, since it had dictated that a charged body, like the electron, should radiate all its energy away as it orbits around the positive nucleus. In other words, according to classical physics, all the atoms in the universe should vanish into energy in a blink of an eye.

Niels Bohr tried to explain why they did not by invoking Planck and Einstein's ideas that energy is always present in discrete amounts called quants. Drawing on this insight, he suggested that in a similar fashion certain discrete—that is, specially quantized—orbits are stable. Werner Heisenberg and Wolfgang Pauli, two young physicists in Arnold Sommerfeld's class at Munich, were troubled by Bohr's picture. Pauli objected that Bohr had grafted the new quantum ideas onto the older concepts of classical orbits. A far more radical approach was needed, he suggested. This approach was provided in 1925, the year Oppenheimer arrived in Europe, by Heisenberg, with his matrix or quantum mechanics.

Heisenberg's approach was abstract and mathematical and required a physical interpretation in order to be properly understood. Max Born took the first step in unfolding the meaning of Heisenberg's theoretical discovery, and later Pauli and Heisenberg explored its implications with Bohr in Copenhagen. In the following year their discussions were fired by Erwin Schrödinger's announcement of what looked like an entirely new and quite different solution to the problem of stable atoms. Schrödinger suggested that electrons were really matter waves that formed stable patterns around the nucleus.

Schrödinger's wave mechanics appeared to provide a more intuitive picture of the atomic world than the abstract model of Heisenberg. Yet the Copenhagen group soon realized that—mathematically speaking, at least—the two approaches were virtually identical. The vital issue remained one of interpretation, which Bohr and his two colleagues continued to hammer out. This Copenhagen interpretation, as it came to be called, eventually became orthodoxy among the physics community.

It was the details of the Copenhagen interpretation that Bohm and Weinberg debated at Berkeley. Hour after hour, day after day, Weinberg hammered away at Bohm, in an effort to convince him of the correctness of the quantum theory and of Bohr's interpretation of its meaning. Each time Bohm resisted, Weinberg would be forced to think up a new example or different explanation, until Bohm acquiesced.

Joe Weinberg enjoyed the battle. He felt that he was leading Bohm toward a clearer understanding of Niels Bohr's philosophy. Bohm, however, found Weinberg's approach harsh and unyielding.[8] Weinberg himself later admitted that, as a young man, he had had "an arrogant bullying style," in which he piled one argument on top of another in an attempt to convince his opponent.[9]

Bohm, for his part, tended to work through flashes of insight. He used feelings, sensations, and inner movements to gain a direct apprehension of what lay beyond ordinary logic. Weinberg's mind was hard, sharp, and clear; he would approach a problem as if he were searching for its weakness, for that place where it could be most advantageously attacked. By way of a counterattack, Bohm would denigrate Weinberg's emphasis on mathematical arguments, calling it Pythagorean mysticism.

One can guess at the sorts of questions that Bohm and Weinberg debated, questions that Bohm was later to reexamine in a particularly radical manner. For example, in formulating quantum mechanics, Heisenberg had arrived at his famous uncertainty principle, which dictates that, at the subatomic level, it is impossible to measure simultaneously both the position and the momentum of a particle with absolute accuracy. (Momentum is the velocity of a particle multiplied by its mass, and one may think of the problem of determining momentum as essentially the same as that of measuring its velocity.)

Heisenberg's uncertainty principle follows directly from the mathematics of his theory. Its meaning, however, calls for an explanation. Position and momentum are the two cornerstones of classical physics. In the large-scale world of planets, billiard balls, and Newtonian apples, the path of an object is defined by its position and momentum. If one knows both the position and the momentum of a comet at a single instant, then one can compute its entire future trajectory.

Admittedly, there are always experimental errors involved in measuring position and momentum, but in principle at least, these errors can be progressively reduced or compensated for by refining the measurements. More important, classical physics takes for granted that a falling apple, a comet, or an orbiting satellite actually possesses a well-defined position and momentum at every instant of its existence.

Heisenberg's uncertainty principle indicated that, at the subatomic level, nature limits the accuracy with which position and momentum can be measured. But if an electron's momentum and position cannot be defined exactly at one instant, then physicists will be unable to compute its path. At one stroke, Heisenberg's principle denied much of the intuitive imagery common to classical physics—the very sort of imagery that the young Bohm liked to employ.

To demonstrate his principle, Heisenberg imagined an actual experiment to measure the exact path of an electron. An ordinary microscope, one that makes use of visible light, would be too gross to determine the position of an electron, but suppose this microscope used gamma rays of exceptionally short wavelength. With this hypothetical microscope, it would be possible to determine the electron's position with a good degree of accuracy. But when a gamma ray photon hits the electron and is reflected back into the microscope, it gives the electron such a jolt that it alters its momentum in an uncontrollable and unpredictable way. The shorter the wavelength of the gamma ray and the more accurately it determines the electron's position, the greater is its unpredictable disturbance of the electron's momentum.

Heisenberg was able to show, by this commonsense argument, that any attempt to measure momentum or position of an elementary particle produces an uncontrollable and unpredictable disturbance. So, he concluded, does nature prevent us from ever knowing the precise value of these variables simultaneously and, consequently, the trajectory of the electron's path.

Niels Bohr rejected the whole tenor of this example. Indeed, so forceful were his objections that they actually reduced Heisenberg to tears. He pointed out that in setting up his argument, Heisenberg had tacitly assumed that the electron actually *has* a path and actually *does* possess a precise position and momentum at each instant. When a measurement is made, it produces an uncontrollable disturbance in one of the two variables.

But Heisenberg was totally wrong, Bohr argued, to begin with this concept of a well-defined, even if in principle unmeasurable, path. He was guilty of confusing new quantum ideas with older classical concepts, such as a path. It is not so much that we disturb the intrinsic properties of an electron by observing it, Bohr argued, but rather that such properties are already ambiguous within the subatomic domain. When Bohm read Bohr's comments, he suggested that Heisenberg's uncertainty principle would be far better called the "ambiguity principle."

Following in Bohr's footsteps, Bohm and Weinberg tried to think up new and ingenious ways of measuring either of two complementary aspects—wave or particle, position or momentum. In each case, they realized, the experimental conditions needed to reveal one aspect proved to be incompatible with its complementary partner.[10]

In their discussions at Copenhagen, Bohr, Heisenberg, and Pauli were seeking an entirely consistent and complete description of the quantum theory, the Copenhagen interpretation. Its staunch defenders, such as Leon Rosenfeld, claimed that it was the only possible interpretation of the quantum theory. It is worth remarking, however, that not even the three protagonists always agreed on what they had achieved. For the rest of their lives, each retained a subtle but importantly different version of the interpretation—which only contributed to the confusion surrounding the quantum theory.

One of the most significant aspects of the Copenhagen interpretation was Bohr's idea of complementarity. Position and momentum, Bohr said, are complementary variables, since the act of measuring or observing one of them precludes the simultaneous measuring or observing of the other. Another aspect of complementarity is the way in which, in some situations, an electron exhibits a wavelike nature, while in others it behaves like a particle. Wave and particle appear as disjoint aspects of the electron, for how can something be both localized and yet spread over all space?

Bohr suggested that no single description can exhaust a quantum system; rather, physics must entertain descriptions that are complementary and, at times, paradoxical. Complementarity became, for Bohr, one of nature's "great truths," something that transcended the particular domain of quantum physics and was necessary in, for example, understanding the nature of consciousness as well. Bohr's colleagues jokingly said, with some justification, that complementarity even extended to Bohr's manner of writing. After setting down one sentence, he would deliberate for a time, seeking a way to state the opposite or to take a contradictory position within the next sentence. This process made for a remarkable degree of obscurity in Bohr's papers, which is probably one reason Bohm and Weinberg were able to argue about Bohr for so long and so intensely.

Although he had not yet admitted defeat in his battle with Weinberg, Bohm could see the beauty in this idea of complementarity. He pictured the individual and society as complementary aspects of something deeper. He also began to think about a more holistic form of consciousness, what he called "the world of light," which was far different from the isolated, individual consciousness that inhabited the world of shadows.

As Bohm argued about quantum theory with his Berkeley friends, he was forced to admit that, when it came to the accuracy of its predictions and the range of the phenomena it encompassed, the theory was amazingly successful. Yet no matter how hard he tried to enter into Bohr's world, he always retained misgivings. Despite the magnificent edifice of thought erected at Copenhagen, some areas of the theory remained unclear to him, and at times Bohr's philosophical explanations appeared to block further investigation. Bohm wanted to know the actuality of what occurred at the quantum level, yet the concept of complementarity, despite its certain beauty, seemed to erect a barrier to further explanation.

CHAPTER 4
From Niels Bohr to Karl Marx

THE PHILOSOPHY OF NIELS BOHR was not Bohm's only influence at Berkeley, for he was also studying Marx and Engels. When he and Joe Weinberg discussed the nature of complementarity, it was not always in terms of the language of Copenhagen. Sometimes they took it as an example of the dialectical struggle of opposites within the natural world described by Marxist theory. At Caltech Bohm had been repelled by the cynicism of his fellow students, and as a result, he had felt increasingly drawn toward socialism. By the time he arrived at Berkeley, he was therefore "in a ripe mood to listen." Oppenheimer's groups proved the ideal nursery for a new political faith.[1]

During his first years in California, Oppenheimer had distinguished himself as a brilliant and cultured academic who took little interest in political affairs. It was even said that he had remained ignorant of the Wall Street crash until several months after the event. But few people could continue to remain indifferent to politics during this period. At the very least, Oppenheimer became aware of the way in which his students were experiencing the effects of the Depression. Some of the Berkeley students joined the International Brigade to fight in the Spanish Civil War, while others got involved in the union movement, teaching and distributing socialist literature.

In 1936 Oppenheimer fell in love with the beautiful but somewhat unstable Jean Tatlock, an active member of the local branch of the Communist party. Under her tutelage, Oppenheimer joined organizations such as the Friends of the Chinese People and the American Committee for Democratic and Intellectual Freedom, met leading

Communists in the California area, and donated funds to organizations associated with the party. He also became friendly with Haakon Chevalier, a deeply committed Communist who was the translator of French novelist André Malraux. Later Chevalier maintained that he and Oppenheimer had been members of the same Communist cell at Berkeley, a claim that Oppenheimer hotly denied.

Oppenheimer's relationship with Tatlock was tempestuous, although at times they came close to marrying. In 1939 she ended the affair, but Oppenheimer continued to support organizations sympathetic to the Communist party and actively discussed Marxist philosophy with his students. In the 1950s Bohm told his Marxist friend Jean-Paul Vigier about the energetic group of Marxists that had been centered around Oppenheimer in the physics department. Vigier was given the impression of an active Communist cell.

In August of the same year, Oppenheimer began an affair with Kathryn Puening, which rapidly led to marriage. Now Oppenheimer saw less of his intellectual friends, and his famous parties became less frequent. While Kathryn remained in contact with Steve Nelson, a Communist party worker, Oppenheimer appeared to distance himself from the left. Later in his life, he claimed that while he had entertained socialist sympathies, he had never been a member of the Communist party and that his interest in Marxism ended at the start of World War II. Others, however, were skeptical of this claim and believed that Oppenheimer remained strongly sympathetic to the Soviet Union and Communism throughout the war and after. In 1994, for example, allegations were made, in *Special Tasks: The Memoirs of an Unwanted Witness—a Soviet Spymaster,* that Oppenheimer, Enrico Fermi, and Niels Bohr actively passed information on the design and construction of atomic weapons to the Russians.[2] Historians and physicists who reviewed or commented on the book discounted the charge, pointing to errors and inconsistencies in the story. Nevertheless, rumors about Oppenheimer and his sympathies have never been fully laid to rest.

Even if Oppenheimer became less public about his left-wing leanings, his students remained enthusiastic about the writings of Marx, Engels, and Lenin. In the 1930s Marxism represented the dream of a better future. Moreover, only Russia seemed to have the will to oppose what many felt was the breakdown of civilization. Although the plight of European Jewry was already apparent, the United States lacked the determination to fight the fascism that threatened them.

In November 1942, with the active support of his friends, Bohm joined the Communist party. The encounter was, however, not particularly dramatic, since he found the party's activities scarcely more interesting than those of the YMHA sports club he had attended back in Wilkes-Barre. "The meetings were interminable, discussing all these interminable attempts to stir up things on the campus which didn't amount to much."[3]

Speaking about his involvement to historian Martin Sherwin, Bohm recalled that the meetings were so boring, he dropped out after only a few weeks. Yet in a statement he later made to the American consul in London, it appears that he remained a party member for nine months. Even after he had left the party, Bohm continued to study Marx, Engels, and Lenin with his friends Weinberg, Peters, and Lomanitz. The more he read, the more excited he became. Marxism, he found, was a truly holistic philosophy, one that embraced culture, politics, and economics yet also gave a special role to science. Its basis lay in dialectic, that dynamic movement in which nothing is fixed and contradictions flow into each other. With his vision of the cosmos as transformation and ceaseless movement, dialectic had a particular appeal for Bohm.

At last Bohm had found something in which he could believe, a structure for his life that was based on reason, science, and the dream of human betterment and freedom. He was part of a movement that would sustain him against cynicism, and he could even share this passion with those around him. If society were organized along Marxist lines, he believed, every action would be rational and taken with regard to the welfare of the whole. Human nature itself would change. Humanity would finally be free of endless cycles of hatred, irrationality, cynicism, and warfare. And if this ideal situation had not yet arisen in the Soviet Union, if the transitional dictatorship of the proletariat had not yet been transformed, it was only because that country was under attack. Given time, Russia would show the world a better society.

For many Jews in this period, Marxism represented a personal salvation and the possibility of global redemption. Writer Henry Roth, for one, described his own involvement this way: "By becoming a communist I could climb out of what I was, negate what I was. And become that kind of a decent person I sought to be."[4]

Deeply committed to Marxism, Bohm was well versed in its maxims when talking to others. He read Engels's *Dialectics of Nature* in the original German, as far as Joe Weinberg could remember. Weinberg was, however, sometimes a little cynical about his friends; admittedly they were all well versed in the Marxist classics, but they tended to pick out passages with which they could sympathize, he thought.[5]

Prior to Bohm's arrival on the Berkeley campus, clouds had already formed on the Marxist horizon. In 1938 physicists Victor Weisskopf and George Placzek returned from an extended visit to Russia and told Oppenheimer of show trials and concentration camps. In this period a variety of charges were being made against the Soviet dictatorship. The general sort of accusation was expressed by a character in Aldous Huxley's novel *After Many a Summer,* published in 1939: "put the abolition of tsardom and capitalism in one scale; and in the other put Stalin, put the secret police, put the famines, put twenty years of hardship for a hundred and fifty million people, put the liquidation of intellectuals and kulags and old Bolsheviks, put the hordes of slaves in prison camps."[6]

The politically faithful were able to discount such stories as the distortions of the capitalist press, or as the tragic but inevitable mistakes made within a rapidly changing society. It was less easy to excuse the mutual defense pact signed between Russia and Nazi Germany in August 1939. Faced with a potential loss of party members, Communists pointed out that even if Stalinist Russia had temporarily been diverted from its revolutionary ideals, Communism itself was not tainted. This was the argument put forward in a 1940 pamphlet "Reports to our Colleagues," which was designed to preserve party solidarity. The pamphlet, circulated by the College Faculties Committee of the Communist party, was widely believed to have been written by Oppenheimer himself.

Following its declaration of war on Nazi Germany, Russia was at last fighting against fascism rather than alongside it. Weinberg expressed the feelings of all his Marxist friends when he said that they suffered every day along with the Russian people in their resistance to the Nazis. They realized the terrible toll that was taken in their defense of Stalingrad: "No one can feel the way we felt. Even when we saw the sham of what was going on in the Soviet Union, of the show trials, we turned our eyes away from them." From now on, Bohm was to be as passionate about Marxism as he had earlier been about physics.[7]

But other events were now to impinge on Bohm's life. The possibility of nuclear fission and the feasibility of developing a nuclear bomb had become clear to physicists by the end of the 1930s. Politicians, however, did not understand the significance of this weapon. Therefore a group of refugee physicists from Europe approached Einstein and asked him to sign a letter to President Roosevelt.

The driving force behind this effort was Leo Szilard, whose experiments at Columbia University in New York City had convinced him of the practicality of an atomic weapon. Einstein had a friendship with Queen Elisabeth of Belgium, Szilard knew, and he hoped that the great man would alert her to the danger of the uranium mines in the Belgian Congo falling into German hands. Accompanied by Eugene Wigner, Szilard made several visits to Einstein, who not only communicated with Queen Elisabeth but agreed to sign the letter, drafted by Szilard, to the American president. Dated August 2, 1939, it did not reach the president until several weeks later, whereupon Roosevelt appointed a three-man advisory committee to look into the matter.[8]

As a result of the committee's report, immediate priority was given to constructing an atomic weapon. On October 8, 1942, Oppenheimer met with Colonel Leslie R. Groves, who was so impressed by the physicist that he urged that he be appointed to head the Manhattan Project, the code name given to the U.S. atomic bomb project. The U.S. government needed an American to oversee the project, someone with sufficient scientific authority to be able to supervise people with international reputations, like Enrico Fermi. Clearly Oppenheimer would fit the bill. It worried the FBI, however, that someone with strong left-wing associations should be in charge of such a sensitive project, and it alerted the Manhattan Project's own security force.

When questioned by the security officers, Oppenheimer was not only open about his past but more than willing to discuss the political activities and sympathies of his students. The great ease with which Oppenheimer named names later shocked his colleagues during the postwar Communist witch-hunts. In Bohm's case, his "father figure" was quite willing to brand him as potentially "dangerous" because of his Communist interests and his susceptibility to Communist influence.

Despite the openness of his disclosures, the security services were still not willing to trust Oppenheimer, and he was placed under surveillance. Surveillance was also maintained on the Communist party headquarters in Alameda County, where its head, Steve Nelson, had been a close friend of Oppenheimer's wife Kathryn. Amongst conversations taped were those that discussed undercover party members and possible sympathizers within Oppenheimer's group at Berkeley.

Although Oppenheimer became indispensable to the Manhattan Project, some officers, like Colonel Boris Pash, chief of counterintelligence for the Ninth Army Corps on the West Coast, and Colonel John Lansdale, a lawyer who acted as Groves's security aide, were never able to trust him. The chief of security at Los Alamos, Captain Peer DeSilva, personally disliked the physicist. On June 12, 1943, Oppenheimer was observed to have spent the night at the house of his former mistress, Jean Tatlock. By the end of that month, Pash was convinced that Oppenheimer was maintaining an active connection with the Communist party and on this ground recommended his dismissal from the bomb project.

Pash also believed that Steve Nelson had organized a cell at the Radiation Laboratory involving Oppenheimer's students. Some of these students, Pash believed, were so sympathetic to the Soviet cause that they would be willing to pass "atomic secrets" to the Russians. Attention now was directed toward David Bohm and his friends.

Any possible sympathy toward the Soviet Union must be seen in its full context. The development of the atomic bomb was the cooperative work of the United States, Canada, France, and Britain, all members of the Allies. But in terms of the sacrifices being made by its troops and civilian population in holding the eastern front, the Soviet Union was a particularly important ally, and if the Germans did indeed develop a bomb, they were as likely to use it against the

Russians as against the British. Oppenheimer urged President Roosevelt, and later President Truman, to agree to an official scientific exchange with the Russians. Since many other scientists agreed with Oppenheimer, it was inevitable that the security services would be concerned. What if someone who believed in the international brotherhood of both science and socialism were prepared to take a shortcut?

According to Rossi Lomanitz, neither he nor his friends ever took this possibility seriously.[9] Bohm certainly knew about the feasibility of an atomic bomb, but at that time he had no idea even of the existence of the Manhattan Project. Lomanitz recalled that Steve Nelson had mentioned that a big weapon was being made but had argued that such things did not win wars. What Russia really wanted from its allies, according to Nelson, was the opening of a second front in Europe.

At all events the Radiation Laboratory was now under close surveillance by security agencies. It was noted that Bohm had been involved in organizing activities for the Communist party and had distributed copies of Earl Browder's *Victory—and After.*[10] (Browder was a trade unionist who rose to become general secretary of the Communist party in the United States. He even ran for president on the Communist party ticket in 1936 and 1940.)

But if Bohm was suspect, the subject of more immediate attention was his friend Rossi Lomanitz. Lomanitz had long wanted to make a dramatic gesture in favor of the left. To this end he organized a branch of the Federation of Architects, Engineers, Chemists, and Technicians (FAECT, Local 25) in the laboratory. Admittedly, it was difficult to conceive why such a union would be needed in the lab, but it seemed the right thing to do at the time.

On August 7, General Groves sent a memo to Harvey H. Bundy, special assistant to the secretary of war; it was delivered by Colonel Lansdale. Enclosed with the memo was a draft of a memo that Groves suggested should be sent directly to the president and a copy of an intelligence report.

In the memo to the president, Groves drew attention to "a considerable increase in unionizing attempts among the scientific personelle [sic]" at the Radiation Laboratory. Several people in the laboratory were "active communists," he said, as well as members of FAECT. Joe Weinberg's name was mentioned, and Groves expressed concern that scientific information could be passed on.[11] Groves suggested that the president should ask Philip Murray, president of the Congress of Industrial Organizations (CIO), to direct FAECT to cease its organizing within the Radiation Laboratory.

Along with this memo Groves sent an intelligence report.[12] The security review was instigated on the oral instructions of Colonel Pash and composed from reports made by Lieutenants James S. Murray and Lyall E. Johnson.

The report notes that FAECT Local 25 is "an organization known to be dominated and controlled by Communist Party members or Communist Party sympathizers" and that it could serve to gather highly sensitive information. It mentions [Irving] David Fox, president of the Radiation Laboratory local, Bernard Peters, a member of the executive committee, and Bohm and Lomanitz as active members.

The union, the report says, follows the party line by giving contributions to Russian war relief and "attempts to bring about the socialization of industry." Reference is also made to Earl Browder, and the numbers of several automobile license plates are included.

Two weeks later, on September 9, 1943, the secretary of war, Henry L. Stimson, forwarded General Groves's draft memo and the intelligence report to the president under his own name. He recommended that FAECT cease its activities at the Berkeley laboratory and be disbanded. In a handwritten addendum he cautioned, "Unless this can be at once stopped, I think the situation very alarming."

On November 2 a letter to the president (from Anna Marie Rosenberg) informed him that, after reading Secretary Stimson's letter, she had spoken with Philip Murray, who would instruct the union to stop its organizing efforts.

Lomanitz had recently been appointed by E. O. Lawrence as liaison between the Radiation Laboratory and Oak Ridge, so it came as a great surprise to him when he was ordered to report to his local draft office. The local draft board assumed that, in view of the importance of his work, the notice was simply a clerical error, and it appealed on his behalf. His draft was deferred until Lomanitz was directed from on high to report to boot camp. He spent the rest of the war years being transferred from camp to camp within the United States, never seeing active duty.

With Lomanitz gone and most of the other scientists seconded for secret work, Bohm was now relatively isolated. He applied to work at Los Alamos but was rejected on the grounds that, having relatives in Europe, he was open to pressure. This official explanation was not all that convincing, since several refugee scientists with relatives in Europe were already engaged in high security work. Bohm probably did not realize that he, Rossi Lomanitz, Joe Weinberg, Bernard Peters, and several others had fallen under suspicion.

With Oppenheimer at Los Alamos, Weinberg was conducting his supervisor's courses on quantum theory and electrodynamics. He worked from notes he had taken during Oppenheimer's earlier lectures. The teaching enterprise was formidable, since each page of his notes was packed with brilliant ideas. Even the transition from one equation to the next could take Weinberg a whole lecture to explain. In time his health broke with the strain of lecturing in the morning, carrying out research in the afternoon, and working on the following day's lecture for much of the night. Bohm now took over the lectures, and he and Weinberg would meet in the evenings to go over the following day's work. Whereas earlier Bohm had had doubts about the foundations of quantum theory, now by teaching it to students he was being forced to engage it directly, using notes that were part of a tradition that went right back to Pauli and the creators of quantum theory itself.[13]

Bohm was reaching the end of his own research project on the collisions of protons and deuterons—only to find it plucked from under his nose. The scattering calculations that he had completed proved useful to the Manhattan Project and were immediately classified. Without security clearance, Bohm was denied access to his own work; not only would he be barred from defending his thesis, he was not even allowed to write his own thesis in the first place! To satisfy the university, Oppenheimer certified that Bohm had successfully completed the research, and he was awarded his Ph.D. in 1943.

The major efforts of the Radiation Laboratory, of which Bohm was still a part, were now directed at obtaining the uranium isotope (U-235) for use in an atomic weapon. Natural uranium contains several isotopes that, in their chemical behavior, are virtually identical. They differ only in the fact that their nuclei contain a different number of neutrons. Most elements are extracted, separated, and purified by a series of chemical reactions, but this method does not work for the separation of the isotopes of a particular element, like uranium, which react chemically in more or less the same way. Berkeley, however, boasted a cyclotron, which E. O. Lawrence had developed to investigate the internal structure of elementary particles.

A cyclotron works by passing a beam of charged atomic particles through a magnetic field, where the particles separate according to the mass of their nuclei. Since the isotopes of uranium have slightly different masses, the thinking at Berkeley went, it should be possible, albeit very slowly, to separate them. The practical difficulty, however, lay in focusing the beams of charged particles. Bohm was now involved in some particularly technical issues surrounding that problem. At first he was discouraged because there were so few theoreticians around with whom he could discuss the work. But in 1943 a contingent of British physicists arrived, under the leadership of H.S.W. Massey. Massey soon had the project organized, and Bohm got involved in the underlying physics. As it turned out, the project would eventually lead Bohm to his first outstanding contribution to physics.

In order to accelerate and direct the uranium atoms in the cyclotron beams, it was first necessary to strip them of some of their outer electrons and leave them with a positive charge. The resulting collection of free electrons and positively charged atoms is known generically as a plasma, a fourth state of matter quite distinct from a solid, liquid, or gas. In a plasma, positively charged nuclei and negatively charged electrons move freely at a temperature of around 20,000 degrees Celsius. Although, on the Earth' surface, such heat can be created only in the artificial conditions of a laboratory, plasmas constitute over ninety-nine percent of the matter of the universe—most stars and interstellar gases exist in this fourth state of matter.

The first theoretical description of a plasma had been given several decades earlier by chemist Irving Langmuir, but as soon as Bohm began his theoretical study of plasmas confined within the extremely high magnetic fields of the cyclotron, he discovered some quite unexpected properties in them. Instead of moving freely, for example, the electrons tended to circle around the magnetic field lines.

By now it was clear to Bohm that the United States was gearing up to produce an atomic bomb and that the work of the Radiation Laboratory was part of this larger program. Bohm, however, did not appear to be concerned about the moral implications of this work. He shared the attitude common to most of the scientists engaged, directly or indirectly, in the Manhattan Project. Chemists, physicists, and engineers working together on any difficult technical and theoretical problem soon became absorbed by the problem. The more exciting and challenging the problem, the more its wider implications are forgotten. In building the atomic bomb, scientists were playing with the very creative forces of the universe itself. They were learning to tame, and then release with enormous explosive power, that which binds together the most fundamental level of matter. The prospect of wielding such a power itself must have had a uniquely seductive effect.

Until the Manhattan Project, American scientists had been regarded as impractical by their fellow citizens—eccentrics in white lab coats who puttered around, engaged in work that had little relevance to everyday life. Now all that changed, as physicists like Oppenheimer and Edward Teller pursued their work with missionary zeal. Even if today they happened to be using atomic energy to build a bomb, they thought, tomorrow they would be using it to build a better world.

In any event Bohm was happy to continue with his research but, as always, only on his own terms. Solving a problem at the purely practical level could never be satisfying to Bohm, since he always had to probe deeper and deeper. As he studied the plasmas he became struck by their extraordinary nature. They began to take on, for him, the qualities of living beings. When physicists studied a plasma by introducing an electrical probe, it would generate a charged sheath around the probe and neutralize its effects. It was as if the plasma were protecting itself and preserving its internal status.

Bohm's Marxism also played a role in his physics. He had thought deeply about the relationship between the individual and society. Marxist society was the rational ideal for humanity, he was convinced, one in which each individual would experience perfect freedom while serving the common good. Capitalism, by contrast, existed only in the hope that the service to the common good would somehow emerge out of a collection of self-motivated individuals. In a perfect society, however, the individual would be an aspect of society and vice versa. Bohm realized that plasma was a perfect metaphor for this society. A plasma functions in a collective way, oscillating as a whole. Yet it is built out of particles, each of which moves freely with its own individual movement. In a plasma, as in an ideal Marxist society, there is a subtle relationship between the free individual and the collective.

Insight was transformed into mathematics as Bohm developed a new theory of the plasma state. When two negatively charged electrons are in total isolation, the interaction between them extends over a long distance. (For that matter, so does the interaction between two positively charged atoms.) But in a plasma an astronomical number of other charged particles rearrange themselves to shield out this long-range interaction. As a result, each of the charged particles in the plasma interacts with its neighbors only over a short range. In this way an electron is able to move in a free, random way, as an individual. Yet the long-range interaction has not entirely vanished; the large numbers of charged particles arrange themselves in such a way as to screen it out. It is the hidden long-range interactions that enable the plasma to behave in a coherent way.

Seen from a distance, a plasma appears to be a series of collective oscillations, involving an astronomical number of particles. Examined at a high magnification, however, only the random motion of individual particles is visible. What Bohm was able to do, in a technical way, was to create a dual mathematical description of the plasma that contains both of these viewpoints. One description (collective coordinates) deals with the collective vibrations, while the other (individual coordinates) explains free individual movement. Yet because the two descriptions are part of a single whole, the collective motion of the whole is enfolded within the random, individual movement, and vice versa.

In a similar way, a person may feel relatively free yet influenced, through common perceptions and meanings, by the overall values of society. Because individuals respond to shared meanings, society as a whole is able to sustain its complex structure. Human societies, like plasmas, are a synthesis of opposites, allowing for both the freedom of the individual and the collectivism of the whole. (Technically speaking, Bohm was, in this way, able to show why plasmas become unstable in an external magnetic field. His theoretical treatment of diffusion in a turbulent plasma has since become known as Bohm diffusion.)

In his research on plasmas, as in all his work, Bohm's thinking was all of a piece. It was not that he was a physicist who happened to have a great interest in politics; rather, politics and physics were, for him, inseparable. The way we look at the world, he reasoned, determines how we act toward it, how we structure our society and derive the tenor of our individual lives. Equally, the society in which we live conditions our values and the way we think about and perceive the world. Science, society, and human consciousness were, for Bohm, all aspects of a greater whole.

Bohm often remarked upon the three great research projects undertaken by the young Einstein. At first glance they belong to very different fields: Brownian motion, which involves the random motion of molecules; the photoelectric effect, which deals with quantum processes; and the special theory of relativity. To Einstein, Bohm pointed out, these three fields were all aspects of one unified perception of physics. In Bohm's own case, his perception went beyond physics and embraced what he felt to be the whole of life.

Although Bohm's work on plasmas was clearly informed by his political views, he was also able to give a complete mathematical description of it. Again, this ability grew out of something deep and intuitive within him, as we have seen during his years at Penn State, when he internalized the working of a gyroscope. That ability to touch preverbal processes at the muscular, sensory level remained with him all his life. It was not so much that Bohm visualized a physical system as that he was able to sense its dynamics within his body: "I had the feeling that internally I could participate in some movement that was the analogy of the thing you are talking about."[14]

To give another example: the spin of an electron is far from the spin of a ball in our everyday world. The ball is able to spin at different speeds, but the electron, spinning about an axis that points vertically up from the floor, can spin only in one of three states—called +1, 0, and -1. Quantum mechanics does, however, allow for linear combinations of these spin states.

Common sense dictates that two spin states—say, spinning in opposite directions around a vertical axis—will combine to produce an intermediary spin around the same upward-pointing axis. Quantum theory, however, indicates that the resultant spin will point in a new direction—spinning around a horizontal axis, for example. The notion is so counterintuitive that physics students either apply the formulae by rote or else visualize the electron in terms of mathematical manipulations of equations, without reference to anything physical. Bohm, however, found that he was able to experience feelings within his body about the way two spin components could combine into something that moves in a new direction.

"I can't really articulate it," he once said. "It had to do with a sense of tensions in the body, the fact that two tensions are in opposite directions and then suddenly feel that there was something else. The spin thing cannot be reduced to classical physics. Two feelings in the mind combine to produce something that is of a different quality. . . . I got the feeling in my own mind of spin up, spin down, that I was spinning up and then down. Then suddenly bringing them together in the x direction (horizontal). . . . It's really hard to get an analogy. It's a kind of transformation that takes place. Essentially I was trying to produce in myself an analogy of that, in my state of being. In a way I'm trying to become an analogy of that—whatever that means."[15] Again Bohm sensed himself as the microcosm of the macrocosm and believed that he contained the laws of nature within his body.

Bohm was now totally engaged in his research, working with great passion. The arrival of the British physicists had encouraged him, and he had made friends with a young physicist, Maurice Wilkins, who was helping to design vacuum pumps used in the laboratory. The two men discussed physics and society, and together they took a trip to the Sierras, where they stayed near a small lake. They remained good friends until Bohm's death. Wilkins later worked at Kings College in the University of London and shared a Nobel Prize with Francis Crick and James Watson for his work on determining the structure of DNA.[16]

On August 6, 1945, an atomic bomb was exploded over

Hiroshima, and three days later a second was detonated over Nagasaki. As reports on their effects came in, scientists were at last made aware of what they had been doing. Richard Feynman became sickened by the very thought of doing physics anymore. On that August day Maurice Wilkins discovered his colleague, Ken Simpson, in a depressed state, saying, "This is Black Monday." Others were not so moved. Wilkins himself was a little ashamed at this, realizing that initially he had not been upset at the news.[17] His reaction was not shared by everyone, for in true *Dr. Strangelove* style, one group immediately began working out the technical details of underground cities that could be inhabited in the event of a nuclear war.

Bohm appears to have retained a somewhat naive optimism about the bomb. Maybe it would be so powerful, he thought, that it would eliminate war and produce international cooperation. Atomic power had enormous potential, and as a virtually unlimited source of power, it could eliminate poverty and lead to peace. True, Rudolph Peierls and Otto Frisch had warned of the biological dangers of nuclear radiation, but Bohm did not appear to draw any implications from this. And if the carnage in Japan had been enormous, he knew, many people had been killed during a single fire-bombing raid.

A month or so after the bombs were dropped, Bohm had a dream. As a child, he used to drive with his father past a ridge of hills containing unfinished concrete buildings. With their ruined look they were nicknamed Concrete City. In the dream Bohm was walking in a valley that could have been in California. On top of a ridge, he saw concrete buildings surrounded by cold blue flames. They resembled laboratories, but their cladding was destroyed and their windows were open, from which emerged a blue fire. It burned but did not consume them. He was depressed within the dream and felt that the buildings represented science that had been ruined.

It is also true that the image of a fire that burns but does not consume would have a very special significance for a Jew. Moses experienced the presence of Jehovah as a voice within a bush that burned but was not consumed. Moses was then charged to lead his people from captivity into a promised land flowing with milk and honey. Bohm too had sought the promised land, a Marxist society guided by a science that both understood and controlled the powers of nature. But now no voice was heard, and the burning bush had become a ruined laboratory.

Following the surrender of the Japanese, the British contingent at the Radiation Laboratory returned home, leaving Bohm isolated for a time. His work, scaled-down now, was directed toward a more advanced particle accelerator called a synchrotron. Bohm's task was to investigate the stability of orbits as the elementary particles circulated around the machine. As soon as Oppenheimer returned to the Berkeley physics department, however, Bohm joined him and went back to fundamental physics.

It is often said that theoretical physicists do their best work when they are young. Maybe youth brings with it a particular clear-headedness and intensity of purpose, but it is also the case that a beginning physicist looks upon each problem in a fresh way, without all the encumbrances of past failures and theoretical cul-de-sacs. Released from his war work, Bohm's creativity flourished. Over the next few years, he tackled all the outstanding problems of his day—superconductivity, the theory of the electron gas in metals, the infinities of quantum field theory, the foundations of quantum theory—with the absolute confidence that in each case he could produce significant insights.

In the curious phenomenon known as superconductivity, an electrical current circulates for years without experiencing any resistance. As yet there was no theoretical explanation for the phenomenon. Bohm spent some time working on developing the theory, then set it aside in favor of the infinities of quantum electrodynamics. (Bohm's habit of setting a piece of work to one side did not mean that he abandoned it, merely that something else had drawn his attention and he would return to it later.)

The theory of quantum mechanics was remarkably successful when it was applied to atoms and molecules. Serious problems arose, however, when it was applied to the electromagnetic field—in a form called quantum electrodynamics. For example, when the energy of an electron surrounded by its own electromagnetic field was calculated, it turned out to be infinite. Even the vacuum of space was calculated to possess infinite energy. Although this conclusion was nonsensical, physicists could find no way of avoiding it.

The problem arose in the following way. An electron is electrically charged and generates an electrical field around itself. In calculating the energy of the electron, physicists have to take account, by means of successive approximations, of the electron's interaction with its own field. When these various contributions are added together, the result is infinite. Many attempts had been made, in vain, to avoid this difficulty. When Bohm looked at the problem, he realized that at each stage of the calculation, at each level of approximation, the wave function that describes the system also became infinite. This should not occur. Bohm suggested that, at each step of the calculation, the wave function should be made regular—that is, "renormalized." If the calculation was done in this way, Bohm argued, the energy of the electron would turn out to be finite.

Bohm explained his approach to Oppenheimer who told him that the problem was not worth pursuing. Bohm felt uneasy at this dismissal, but having great admiration for the man, he could not bring himself to question his judgment and dropped the topic.[18]

Despite Oppenheimer's damping effect, Bohm wrote a scientific paper on his solution. But the referee's report was negative, he felt, so he did not bother to revise it for publication. Later he learned that Pauli had refereed the article and that he should not have taken his critical comments so seriously. Joe Weinberg was annoyed at Bohm for abandoning work that he considered important; in fact, it contained the essence of the renormalization theory, which later came to dominate theoretical physics.

Bohm did, however, circulate an abstract of the article, which came to the attention of John Wheeler, who had been Einstein's assistant. The physicist was sufficiently impressed to offer Bohm a position at Princeton. The offer was attractive, and Bohm made up his mind as soon as he learned that Oppenheimer would be head of the Princeton Institute for Advanced Study.

CHAPTER 5
Princeton

BEFORE HIS APPOINTMENT at Princeton officially began, Bohm was invited by the physics department there to give a seminar on his work. In the audience was a young physicist, Eugene Gross, who was in search of a thesis supervisor. During Bohm's lecture Gross not only heard that plasmas are the stuff of electrical arcs and stellar atmospheres, but he was introduced to the bold speculation that the astronomical number of freely moving electrons in a metal could also be treated as a plasma. In most substances electrons remain tightly bound to their respective atoms, but in a metal atoms arrange themselves in a regular lattice pattern, leaving their outer electrons relatively free. Moving through the lattice, these electrons were thought, by theoretical physicists in the early decades of this century, to behave like a gas. Bohm now proposed that this "electron gas" must also have a plasmalike nature. His talk opened up a new world for Gross, who wrote up his lecture notes and showed them to Bohm. Bohm agreed to take him as a student.[1]

Bohm's first impressions of Princeton were not particularly positive. The university was situated in an industrialized region, and the physics department itself was not as stimulating as he had hoped. He saw John Wheeler only occasionally, and he did not get along well with Eugene Wigner, a distinguished and exceptionally talented physicist of the generation of Heisenberg and Pauli.

Apart from his stint filling in for Joe Weinberg at Berkeley, this was Bohm's first teaching position. At first his style was far from effective. He delivered his lectures in a low mumble, mainly directed at the blackboard, which made it difficult for anyone beyond the first few rows to understand. He also tended to stand in front of the blackboard, blocking his students' view as he tentatively wrote down and then erased equations.[2] Bohm was not very happy teaching undergraduate courses, and in the following year he insisted on giving a graduate course in quantum theory.

Physicist Kenneth Ford later remembered this course.[3] By this time, Bohm's teaching style had improved. There were about a dozen students in the room, and Ford found the lectures inspiring. Not only did Bohm teach the usual problem-solving techniques, he probed deeply into the meaning of quantum theory itself. His intense, absorbing teaching style gave the class a sense of unity, in which everyone felt at ease and able to voice their questions. Although Bohm was shy, his students found him more approachable than the other Princeton professors, who maintained artificial barriers of authority and position.

In addition to teaching, Bohm was supervising Ph.D. students, including Marvin Weinstein, who had come with him from Berkeley, and Eugene Gross.[4] Gross was happy to be doing physics that was deep, speculative, and original rather than working along well-worn paths. He did, however, have to adjust to Bohm's method of doing research—taking long walks with breaks for coffee, then returning to the office, at which time the mathematics would fall into place.

Unlike other professors, Bohm was noncompetitive in his approach. Once, after Bohm had spent time developing an aspect of his plasma theory, Gross came across a paper on the same topic by the great Russian physicist Lev Landau. (In the aftermath of war, the publication of scientific journals was often delayed by as much as a year.) Bohm was not at all perturbed to learn that the Russian had preceded them with his theory of Landau Damping; rather, he admired the elegance and incisiveness of Landau's approach. In Gross's opinion, "Dave's essential being was then, and still is, totally engaged in the calm but passionate search into the nature of things. He can only be characterized as a secular saint. He is totally free from guile and competitiveness, and it would be easy to take advantage of him. His students and friends, mostly younger than he is, felt a powerful urge to protect such a precocious being."[5]

In addition to sharing scientific research, Bohm and Gross also discussed philosophy. Gross had taken a course from Alonzo Church, a Princeton professor of mathematics who specialized in symbolic logic, and he began to explore dialectical modes of thinking with Bohm. Gross, a Catholic, discussed with Bohm the nature of the Holy Trinity, remarking that when it came to intangibles, Catholic theology resembled Bohr's Copenhagen interpretation.

As Bohm talked, he wove vast structures of the imagination. Once during a party, Gross recalled, his supervisor created a convincing theory of the existence of ghosts and devils.[6] Yet in this very skill Gross sensed danger; so much did Bohm love talking and explaining things that he became disinclined to get down to what Gross felt was the real business of physics—that is, making detailed calculations.[7]

With his other student, Marvin Weinstein, Bohm went back to the problems of superconductivity and infinities in field theory, albeit this time from a new direction. Approaching the latter problem, he wondered whether the electron has a finite size, or if space itself can only be defined down to a certain minimum size. This approach certainly eliminated those infinities, yet the more Bohm thought about it, the more his approach seemed arbitrary. Other physicists would have gone ahead and published, but Bohm paused and asked even deeper questions about the meaning of space and time and its role in relativity and quantum theory. Solving isolated problems was not enough for him; he felt the need to look at the whole of physics and develop a more radical theory.

Already Bohm was beginning to cut himself off from the mainstream of physics. Unlike most people around him, he could not accept that great advances could be made only by individual "geniuses"; rather, he believed that each person has access to a great well of creativity. Deciding to wrestle with the deepest problems on his own, Bohm went over to the Institute for Advanced Study to talk things over with Oppenheimer. His former mentor did not seem particularly interested and suggested that Bohm follow the mainstream. In an effort to be helpful, Victor Weisskopf took Bohm to one side for a heart-to-heart talk. There were times, he said, when Oppenheimer and the others wondered if Bohm could be a little mad. Why did he not get down, Weisskopf asked, to solving real problems that would make his scientific reputation? Irritated, Bohm replied that he was not going to spend the rest of his life masochistically churning out calculations. Although Bohm remained unconvinced, Weisskopf's remarks stung him deeply, particularly Weisskopf's own assertion, "in a very virtuous way," that "I [Weisskopf] am working on much more modest things." Bohm took this as an "immodest modesty." Rather than focusing on physics, Weisskopf and his colleagues were really focusing on the limits of their abilities. Weisskopf was "a namby-pamby" with no courage, Bohm concluded, "a person who adopts opinions because he feels they are relatively safe."[8]

In short, Bohm was supposed to "know his place." It was all right, he was being told, for that tiny handful of "geniuses" to do what they wanted, but everyone else had to get on with good solid work. Even when the young Richard Feynman announced his Feynman diagram method (which later became one of physics' most important tools), he was said to be an upstart who should not be talking nonsense in front of his elders. Doing physics had become a matter of either jumping on a bandwagon or working on a problem that was already being tackled by a major physicist. Looking back, Bohm realized that things had not been too different at Berkeley, where he and his colleagues were all assumed to be little people compared with Oppenheimer. John von Neumann had suggested that physics is organized like a church, with a pope, cardinals, and bishops—Bohr was pope, and no doubt, as Bohm pointed out, von Neumann fancied himself one of the cardinals.[9]

The world Bohm had decided to enter was one that very few people ever encounter, a world inhabited by only the best scientists, composers, artists, writers, and poets. The mastery of technical skills was, of course, a necessary condition for entry, but a person always had to have something more as well: a combination of deep physical intuition and unbounded creativity. Above all, Bohm was possessed of an extraordinary energy and focus. While Oppenheimer excelled in the wide range of talents and in knowledge of physics, he lacked that essential single-mindedness that enabled Bohm to hold an idea vividly within his mind for hour after hour.

In the world of his imagination Bohm traveled freely, erecting great abstract structures of thought. He was like a painter who sketches out a portrait, then gives attention to certain features, the shape of the mouth, the sheen on the hair, the folds of a dress—yet all the time maintaining an instinctive grasp of the whole work, its balance and impact. Like a painting, Bohm's theoretical inventions were free creations yet were constrained by internal laws of consistency. Their overall form was guided by his deep intuition, and by a constant correspondence between theory and the world of observation and experiment.

For Bohm, physics was an inner journey grounded in the conviction that his own body was a microcosm of the universe. In his everyday life he appeared uneasy with his body and, at times, blocked from his inner feelings. Yet like the sightless person whose other senses are heightened, Bohm may have been sensitive to certain aspects of his body at some other level. He was aware, for example, that the pains of his childhood remained fixed and knotted in regions of his being.

Bohm would often sit slumped in a chair and, particularly during the last decades of his life, appear pallid and drained of energy. Then as he began to talk, a passion would rise in him. His face would animate, and his body would make sudden movements, most remarkably the fluttering of his hands. Subtle internal movements, body dispositions, and muscular tensions could be related to ideas in physics, Bohm knew, as if the order of nature were contained within his own body. As he talked, one sensed that his body movements were the outward manifestation of unfolding thoughts and intuitions.

Kenneth Ford, a graduate student, remembered him wandering up and down the halls, looking abstracted and jangling a bunch of coins from hand to hand.[10] Daniel M. Lipkin, who used an office across from Bohm, would hear that same noise coming from the physicist's darkened room at night. It was the "unmistakable sound of a very large collection of coins raining down into his hands after he had evidently thrown them up into the air. There would be a long gap as if he were thinking deeply, then they would be thrown again." Intrigued, Lipkin spied on Bohm and watched as "he cupped both hands to form a bowl, palms up and little fingers touching or crossed, and held a very large pocketful of coins, which he would throw straight up, perhaps six inches to a foot free of his hands; and he would then catch the coins as they fell back down, using only his cupped hands." Lipkin marveled at Bohm's accuracy, and he felt that what he was doing went beyond relaxation and in some way may have helped him to "dwell on the deep mysteries of gravitation (and maybe also of probability?) and, by extension, of life in general."[11] During this period, Bohm decided once and for all that he would carve his own path in physics.

At the time he was living in a large room on Nassau Street with two double beds, one of which he rented to Kenneth Ford. Ford saw little of Bohm, who seemed so totally wrapped up in physics that he did not go to parties or visit the Nassau Tavern with other members of the department. Bohm did, however, enjoy movies, and most evenings he went to the cinema after dinner, then returned to the physics department, where he would work until the small hours. When Ford left their room in the morning, Bohm would still be sleeping. Bohm's odd hours, combined with his retiring nature, gave Ford a feeling of isolation, and after one semester, he moved into the graduate college.

As the months went by, Bohm was gradually drawn into a group of friends with whom he felt more relaxed. They had in common a passion for talking and dreaming of a new society. In their company, Bohm could express his ideas and even drink and enjoy himself. Bohm had not been antisocial, merely shy with strangers and uneasy with small talk. Given an audience interested in physics or politics, he would blossom and talk long into the night.

One of the first of his Princeton friends was George Yevick, a physicist with a mercurial personality and childlike enthusiasm. Bohm, dejected by the cynicism around him that treated ideas as commodities that could bring prestige and money, found Yevick with all his enthusiasm to be a breath of fresh air. If Yevick's political views did not swing as much to the Left as Bohm's ("how can such a nice man [Yevick] have such dreadful politics?" one of Bohm's friends asked), at least they could dream together of creating a "scientist party" that would rule the world. The fact that people like Yevick existed gave Bohm hope that passion and innocence could survive in a corrupt world.[12]

As his circle of friends widened, only one thing was missing in his life, but this lacuna too was filled several months after he met George Yevick. One day in January 1948, George's wife Miriam joined a group of friends who were talking with a young physicist outside Fine Hall. The young man was introduced to her as David Bohm, and she later described the encounter as a "*coup de foudre*."[13]

A few weeks later, Bohm was invited to dinner at the Yevicks', and at the end of the evening Miriam walked him to the bus stop. Bohm later told her, "That was when I knew you were interested in me."[14] Miriam began to visit him at his office in the evenings, but despite his attraction to her, Bohm agonized about deceiving his friend. In the end Miriam spent the night with him, which caused Bohm to worry that his landlady might catch them together—he persuaded Miriam to leave first thing in the morning!

Their affair continued, and Miriam took a job at the National Bureau of Standards in Washington. Bohm would now travel up to Washington and they took a room together, although he was nervous that the owner would demand to see their marriage certificate.[15]

Through Roy Britten, a left-wing physicist, Bohm met another group of friends. Britten rented a farmhouse near Hightstown, New Jersey, some ten miles outside Princeton. Dissatisfied with his Nassau Street room, Bohm moved in with the Brittens. David Pines, who had been Bohm's student at Berkeley and was now working with him at Princeton, also lived at the farm with his wife Suzy. In their company Bohm would walk in the country and enjoy parties attended not only by physicists but by singer Pete Seeger and several painters.[16]

A community of artists and writers lived in the nearby town of Roosevelt, they learned. It had been founded by the painter Ben Shahn and his wife, originally as a Jewish religious community. Bohm and his friends made contact with the group, and the result was a new series of friendships and parties.

With Bohm at the farm, his meetings with Miriam Yevick became easier. She was delighted that he was participating in the various social gatherings that surrounded his new life. Those who knew Bohm in later life, and who were struck by his abstinence and austerity, would have been surprised to see him urge Miriam, who didn't particularly like alcohol, to accept a drink. Even if she did not enjoy the taste, he told her, it was the sociable thing to do.

Miriam Yevick was a mathematician who could appreciate the meaning of Bohm's work. Their relationship was therefore satisfying to him both intellectually and sexually. While lying in bed, he would explain to her his ideas about the electron or the nature of the dialectic. Sometimes he felt compelled to get up and go for a walk, then on his return, he would explain the new insights he had conceived. On other occasions he would tell her that he must rest; sleep was of great importance to him because of the amount of thinking he had to do!

Miriam was a warm, effervescent person, able to counterbalance Bohm's tendency to anxiety, worry, and depression. Yet their illicit relationship turned out far from ideal. Bohm clearly felt guilty about what he was doing, yet he seemed unclear as to how much George knew of the affair, or to what extent he tacitly condoned it. Complicating the situation was the fact that "free love" was part and parcel of belonging to the left in the 1940s and 1950s, when Wilhelm Reich's theories on sexual repression were well known. (Some years later Bohm talked knowledgeably about Reich to his student Christopher Philippidis).[17] Reich, who gave great importance to the orgasm, proposed that sexual repression had played a role in maintaining fascist states. Clearly, the left believed, any social transformation would also include a liberation of the sexual impulse—although this did not necessarily mean sleeping with one's best friend's wife.

The friends among whom Bohm and the Yevicks moved dreamed of the transformation of society and, along with it, the decay of outmoded, bourgeois institutions. In theory, at least, it was perfectly acceptable for affairs to take place that transcended the boundaries of marriage, and the excluded partner would even pretend ignorance. George, Miriam, and David were playing such a game. Yet Bohm could not leave the situation alone and was constantly agonizing about their triangle. When Miriam raised the possibility that the two of them should live together on a permanent basis, Bohm was indecisive, which made her angry. Rather than sweep her off her feet and ask her to leave her husband, he seemed happier to debate both sides of the issue. As a result, their relationship became tempestuous, alternating between breakups and reconciliations.[18]

In this period Bohm also got to know Melba Phillips, a physicist who had been Oppenheimer's first research assistant and who lived with a group of friends in New York City. Phillips was sympathetic to Bohm's political views. She had a special affection for him and tried to make her home a social haven for him, including visits to friends in Hunderton County, New Jersey. Bohm would arrive for the weekend, sometimes traveling to the nearby town to attend a Gilbert and Sullivan opera but mostly talking late into the night with Melba's friends about physics and left-wing politics.[19]

Phillips noticed that Bohm was excessively preoccupied with his health. Wherever he was, he insisted on taking his afternoon nap, sometimes bringing along an eye mask to block out light. On one afternoon a large poplar tree crashed down very near the house, causing a great deal of noise and some damage. Worried that someone could be crushed under the tree, a search was made. Everyone was accounted for except Bohm. Only after what seemed like a very long time was it discovered that he had slept through the whole event.[20]

Phillips thought that Bohm possessed two talents: a talent for being unhappy, and a talent for getting people to want to take care of him. Despite his inner sadness, however, photographs taken in the late 1940s show him to be handsome, slim, and cheerful. Women were often fascinated by him, and as Miriam Yevick noted, he would focus on a person with such intensity as he talked that a young woman could easily become bowled over, convinced that he was giving her special attention.

As a complement to his active social life, Bohm's work in physics was also going well. His research on plasmas had moved ahead, first with Eugene Gross and then with David Pines. The initial work with Gross had used the approach of classical physics, but now he and Pines were investigating plasmas in metals from the quantum mechanical point of view. The papers they would publish together established some of the most significant ideas in solid-state physics.

Bohm had largely rejected what he felt were the narrow goals of the other professors in the physics department. He preferred to spend his time with graduate students, who he found more enthusiastic and more open to new ideas. During his second year Kenneth Ford prepared a short paper for the department's Journal Club. It focused on new experimental results from Oak Ridge, involving the scattering of low-energy neutrons from various atomic nuclei. Ford made use of what he had learned in Bohm's quantum theory course to make some calculations based on a very simple model: this involved ignoring all the fine details of attractive forces in the nucleus and assuming that the neutron was located in a "square well."[21]

His talk lasted only twenty minutes, but it generated some excitement since theoretically Ford's calculations should not have been in such good agreement with the experiments. Something curious was going on. When Bohm discussed the results with Ford, he helped him to fill in the background details and suggested additional ideas. For several weeks the two men worked together, and the result was a letter to the journal *Physical Review*.[22] It was Kenneth Ford's first published scientific work and, he later felt, one of the most significant papers he ever published.

A colleague of Bohm's later years, Basil Hiley, said that in his scientific work, his friend was like a helix—he moved forward by constantly circling around.[23] A writer or a composer may spend many months or even years on a particular work, building, refining, and polishing, until finally it can be set aside for the next project. Bohm worked in a very different way. He would discover some aspect of physics that deeply interested him, and it would take up all his attention for weeks or months. He would probe into the problem with considerable focus, until one day he appeared to lose interest and move on to something new. Yet he never totally forgot the old problem—he appeared still to be working in the back of his mind. Perhaps a year or two later, Bohm would go back to the old problem with a new perception or some new knowledge, and he would push forward again. In this way he could sustain a particular effort for years, even decades, often transforming it beyond all recognition. Indeed, at the end of his life, he was still returning to problems that he had first thought about at Berkeley and Princeton. In some ways Bohm's approach evoked that of his favorite painter, Paul Cézanne, who tended to work on a number of canvases over many months, constantly returning, revising, adding, and taking away. Perhaps it is no accident that both men were seeking to define new structures, one in painting and the other in physics. For both, adding something new in one area seemed to induce an immediate revision of everything else.[24]

With another graduate student, Tod Staver, Bohm took a new look at the problem of superconductivity. This time he could draw upon the techniques of collective variables that he had developed for treating the plasma. His idea was to create a theory about how the electron plasma interacts with the metal lattice itself. In essence his approach anticipated the work of Herbert Fröchlich, John Bardeen, Leon N. Cooper, and J. R. Schrieffer, who later established a basic understanding of the phenomenon of superconductivity. Unfortunately, Bohm never took his own work to its conclusion. He was still pursuing the approach when he left for Brazil. Staver himself was killed while skiing in Massachusetts, and David Pines later wrote up Staver's thesis work.

Another graduate student who enjoyed his opportunity to discuss physics with David Bohm was Sam Schweber. Schweber found his own supervisor, Eugene Wigner, to be stiff and formal, so he spent most of his time with Bohm. The two men shared similar political views, which helped cement their friendship. They would meet during the evenings and on weekends and exchange ideas in a highly active way. One evening Schweber, along with Eugene and Sonia Gross, stayed up the whole night listening to Bohm talk about physics, philosophy, sociology, and politics.[25]

Schweber also attended Bohm's lectures on advanced quantum theory. Bohm took great pains to make his lectures lucid and accessible, pointing out the limitations of each particular approximation that he was using. By now he had decided to do more than a simple lectures series. He was determined to develop a book, to be called *Quantum Theory,* partly because of his distaste for the direction in which mainstream physics was moving. During the golden age of Planck, Bohr, and Einstein, it seemed to him, physicists had always been deeply interested in the physical meaning of their theories and their philosophical underpinnings. Understanding, in those days, was more important than problem-solving or predicting some new phenomenon. Since that time, Bohm believed, physics had diminished, and individual physicists were more interested in displaying their mathematical skills than in using their intuition.

Swimming against the tide, Bohm intended to write a book in which mathematics would be given its proper place—as a tool guided by intuition and physical understanding. Quantum theory would arise out of the necessity of understanding new phenomena and experiments. Physics would take pride of place, and mathematics would be used only to demonstrate results that had already been arrived at through an underlying physical understanding. In addition, Bohm would discuss the philosophical aspects of quantum theory, pointing out what was new and surprising about it. As his lecture series progressed, Bohm worked them into a more coherent written form: the manuscript was distributed in lavender-colored dittoed reproduction.[26]

Oppenheimer felt that Bohm's new interest was a waste of time. When the book was finished, he remarked to a colleague, the best thing Bohm could do would be to dig a hole and bury it. Bohm was hurt by the remark. Oppenheimer seemed to believe that each physicist should be content to supply only a tiny brick toward the edifice of truth. Clearly Bohm's book was not considered one of those tiny bricks!

While he was writing his book, Bohm moved to a room at One Evelyn Place in Princeton. The change of location was to be particularly significant since the daughter of the house's owners eventually became his fiancée. One Evelyn Place, owned by Erich and Lilly Kahler, had become a mecca for visiting Europeans like Thomas Mann, Hermann Broch, and Wolfgang Pauli. Lilly, who came from Vienna, was a remarkable woman, somewhat like Alma Mahler in her ability to attract men of exceptional intellect. Not only had she been close to Einstein (it is reasonable to suppose that she had an affair with him), she was attracted to Erich Kahler, a German philosopher and student of cultural history. Kahler was the author of such influential books as *Man the Measure*, *The Transformation of Man*, *Der Beruf der Wissenschaft* (The Profession of Science), and *Der deutsche Charakter in der Geschichte Europas* (The German Character in the History of Europe). An idea of his circle can be gleaned from the little book *Erich Kahler*, which his friends presented to him as a gift—each one had contributed an essay, poem, letter, or drawing; the overall effect is that of a microcosm of intellectual life in the first half of the twentieth century.

Erich Kahler had arrived in Princeton, apparently at the urging of Thomas Mann, in 1939, a penniless Jewish refugee. When he learned of a house for sale on Evelyn Place that would require raising a down payment of $10,000, Einstein offered to lend him the money. Kahler pointed out that $10,000 was a large sum and said he would certainly offer Einstein a good rate of interest, to which the physicist replied, "Am I a Shylock?"[27]

One Evelyn Place was filled with music, philosophy, science, and the arts. Einstein would drop in to play his violin or talk with Thomas Mann and Jacob Bronowski. While Bohm lived there, he was steeped in an atmosphere of exceptional expatriate minds constantly exploring new ideas. His new location also introduced him to Lilly's daughter, Hanna Loewy, a particularly beautiful and talented young woman. She had studied philosophy and assisted Erich Kahler with his books, translating poetry from the German and offering critical opinions. In *The Tower and the Abyss: An Inquiry into the Transformation of the Individual,* Kahler wrote:

> The completion of this book has been furthered and encouraged by the aid of friends to whom I have every reason to be profoundly grateful.
>
> First amongst these is Miss Hanna M. Loewy who not only helped me in all the various operations which the finishing of a book involves, such as revising the manuscript, selecting and pondering material of documentation, checking references and so forth, but beyond all these special tasks took the book in her comprehensive care, devoting to it the unusual gifts of her mind. She also contributed some fine translations.

In addition to philosophy Hanna was interested in the arts and planned a career for herself in documentary filmmaking. At the time Bohm moved into the Kahler home, Hanna was spending most of her time with artists and filmmakers in New York. On her periodic visits home, however, an attraction grew up between them, and soon they became lovers.

Hanna also helped Bohm with his book, particularly the final chapters, in which he discussed some of the most difficult and subtle arguments about quantum theory's interpretation. Friends like Melba Phillips and David Pines had been able to comment on the physics, but Loewy had been trained in philosophy, and so she talked over these later sections with Bohm, making changes, attempting new approaches, rewriting and editing.[28]

The Copenhagen group of Bohr, Heisenberg, and Pauli had argued that quantum theory gave a complete account of the microworld, so that no additional assumptions or hypotheses were required. Einstein, however, had always felt that the theory was incomplete and provisional, being merely an intermediate step on the way toward a deeper theory.

Einstein's major objection to the quantum theory was its lack of an objective description. Another problem, he felt, was its probabilistic character; hence his famous objection that "God does not play dice with the world." When referring to games of chance, *probability* is another word for our ignorance of the many tiny perturbations that act on a tossed penny or on a ball buffeted in a spinning roulette wheel. The outcome of the event is totally causal and determined, but because we cannot compute exactly the effects of all its causes, we must resort to probability, knowing only the chance that a penny will come down heads or that a particular number will come up on the roulette wheel. In quantum theory, however, probability is not a measure of ignorance of a complex underlying, causal process; rather, it is the absolute and irreducible manifestation of the quantum world.

To take an example, when an electron collides with another elementary particle, quantum theory is able to predict only the probability that we can then find that electron in a given region of space, but cannot say for sure. The value of the probability is computed from the wave function of that electron. Quantum predictions are purely probabilistic, since no matter how carefully physicists control and measure the system, these probabilities can never be reduced to certainties. This was something that Einstein could never accept.

To be sure, the situation is somewhat paradoxical: if a Geiger counter were placed in a particular region of space, after all, it would either register the arrival of an electron or it would not. In other words, it would give a precise answer. That is, although only probabilistic predictions are possible in quantum theory, every observation gives a definite result. How is it that the wave function, which expresses these various probabilities, suddenly "collapses" into a well-defined result? This question is one version of the quantum measurement problem, which Bohm was attempting to explain in the final chapter of his book.

Niels Bohr had attempted to answer the question once and for all, but his solution did not convince Einstein. Other physicists made suggestions, too. Some believed, with Einstein, that the theory was incomplete and required the addition of some new factor to account for the discontinuity between probabilistic predictions and definite outcomes. Others hypothesized that the theory applied only to the atomic domain and lost its validity at the scale of a Geiger counter. As time went on, even more bizarre explanations surfaced, such as speculations about the action of the consciousness of the human observer on quantum systems; or that each of a host of possible outcomes did in fact occur but in quite different universes.

Niels Bohr and his Copenhagen colleagues had tried to silence such speculations with a single coherent account, one that did not resort to metaphysics. Yet as we have seen, the Copenhagen adherents did not always agree among themselves. Heisenberg, for example, used Aristotelian language to suggest that during a quantum measurement a multiplicity of *potentialities* is transformed into a single *actuality*, while Wolfgang Pauli, drawing upon his Jungian interests, argued that the problem was evidence of the irrational in matter.

The Copenhagen explanations were subtle and difficult to follow. Bohm's goal, for his book, was to present them in language that was simple, clear, and concise. It was a Herculean task, since it was not entirely clear that Niels Bohr's arguments themselves were, in fact, totally consistent—at some deeper level there might be confusions and inconsistencies. In any event, with the help of the critical reactions of his friends, Bohm attempted to present Bohr in the best possible light.

At one point Bohm gave a seminar on the nature of the quantum theory. The hall was packed with some of the great names of physics, including Albert Einstein, Hermann Weyl, John von Neumann, Eugene Wigner, and J. Robert Oppenheimer. Bohm's talk gave rise to a heated debate, yet he dealt with the various objections that were raised in a calm and confident way, Sam Schweber remembered. At that moment Schweber realized that doing theoretical physics required not only intuition and intelligence but also a degree of courage.[29]

Finally Bohm's book, *Quantum Theory,* was finished, and Eugene Wigner gave a party—even Bohm's father attended. This celebration was also the occasion when Hanna Loewy announced her engagement to Bohm. Their engagement may have been formalized yet things were not well between them. Hanna was a warm, beautiful, and exuberant woman who could sweep a man off his feet—in fact, she boasted of her other lovers while she was engaged to Bohm. She was well accustomed to powerful intellectual men—after all, Einstein had been a regular visitor at her home. Bohm always had something special, something appealing to her, yet in the last analysis she realized that she did not really love him and would not be able to live with him.[30]

Although Bohm's major passions were of the mind, he did at times experience strong sexual urges that demanded immediate satisfaction. "In bed he was good at the male macho stuff," Loewy put it, "but there was little eros in it."[31] Among those who devote themselves to a life of the mind, there is always a danger that sexual energy will be raw and undifferentiated, carrying the person in its grip with strong impulses and urges. This discharge of sexual energy was something Bohm needed, yet when it happened, it always somehow disappointed him. He even complained to Miriam Yevick that the act of love should have had a transcendental conclusion.[32]

Even in bed Bohm's physical reserve and awkwardness would not leave him, and he was unable to touch Hanna in a sensual way. Her intuition was that Bohm had become cut off from feeling and sensation almost to the point of autism. He would never be able to enter her world of sensation, she realized, and enjoy food, art, music, and sex as she did.[33] Bohm himself was well aware of his problem and later wrote of the "blemishes on my character . . . sometimes I feel as if the space between my intestines and my lungs is one big emptiness, and that instead of having sincere feelings and desires originating there, it is all action by habit and hypocrisy."[34]

CHAPTER 6

Un-American Activities

On the evening of December 4, 1949, a U.S. marshal entered Bohm's office at Princeton and arrested him on the charge of contempt of Congress. The trauma associated with the events that unfolded never left him—shortly before his death, as we shall see, he relived it in a particularly painful way.

The circumstances that led up to his arrest had been set in motion back at Berkeley, when he and his friends had met to discuss Marx. After the war against the Axis powers ended in 1945, the United States became involved in the cold war, and with it rose a new anti-Communist hysteria. The disorder soon to be known as McCarthyism swept the United States, but the use of this name places all blame for the phenomenon at that particular senator's feet. Many other groups and individuals had already been persecuting Communists since 1945. In fact, the term *McCarthyism* only came into use a year after Bohm was called on to testify.

During the war the security services had believed that a Communist cell was operating at the Radiation Laboratory at Berkeley. Later, when Oppenheimer was questioned by Captain Peer DeSilva, chief of security at Los Alamos, he had talked about his students David Bohm, Rossi Lomanitz, Bernard Peters, and Joe Weinberg as potentially dangerous. In a memorandum dated January 6, 1944, DeSilva recorded that Oppenheimer believed some of them were "truly dangerous." Bernard Peters, he felt, was dangerous because he was a "crazy person" and "quite a Red," so that his actions were unpredictable. Bohm's dangerousness lay in the possibility that he would be influenced by others.

Army intelligence had long suspected that someone in the Radiation Laboratory had passed information to Steve Nelson. According to its agents, a certain "Scientist X" had met with Nelson and read him "a complicated formula," which Nelson copied down. Nelson had then arranged a meeting with the Soviet vice consul, Peter Ivanov, and was later handed "10 bills of unknown denomination" by the consulate's third secretary. But who was "Scientist X"? The investigation produced a surveillance photograph of four men going to a meeting on Van Ness Avenue. These figures, they believed, were Joe Weinberg, Rossi Lomanitz, Max Friedman, and possibly David Bohm. Lomanitz, Bohm, and Weinberg had been placed under constant investigation. Reports noted that they were frequently together—not an unusual occurrence for students working under the same supervisor. Suspicion initially focused on Lomanitz and later on Weinberg as the putative spy.

Now well established at Princeton, Bohm could have been excused for assuming that all this was water under the bridge. His friend Rossi Lomanitz, however, was not so lucky. One evening in 1946 Lomanitz and his partner, Mary, were preparing dinner when two federal agents arrived and took him to their headquarters, where they questioned him for several hours. They conducted the interview along traditional good-cop-bad-cop lines, the tough guy speaking of "commie bastards" and the nice guy asking vague questions about people's reputations.

Mary became uneasy at the length of time Rossi had been gone, and having taken down the license number of the agents' car, she called the Berkeley police. During his interrogation Lomanitz heard the phone ring in another room, and shortly afterward he was driven home with an ominous "We'll see you again." From that time on he was followed, he realized, and sometimes even filmed.[1]

By now the House Committee on Un-American Activities (HCUA) was holding a series of hearings, in part to determine what had occurred in the Radiation Laboratory during the war. On April 21, 1949, a subpoena was served on Bohm at his office, requiring him to testify.[2] Lomanitz had also been subpoenaed, and the two friends began to correspond. Bohm discussed with his friends what course of action he should take. At the best of times Bohm was inclined to anxiety, worrying about the possible outcomes of even the simplest decision. The situation he now faced was particularly serious. Was there a way he could avoid the subpoena? If he were called on to testify, how could he avoid compromising his friends?

Einstein advised that he refuse to participate outright. To appear before the committee, he said, would validate the hearings. In giving this advice he warned Bohm, "You may have to sit for a while," meaning that he might have to go to jail.[3]

In the end Bohm decided that he would attend but refuse to answer specific questions. The grounds for this refusal would be a combination of the First and Fifth Amendments to the Constitution. The former, which guarantees the right to freedom of speech, had been invoked by Oppenheimer's brother, Frank, who held that it also established the right not to say anything. The Fifth Amendment guaranteed the right against self-incrimination.

But what protection these amendments offered a witness testifying before the HCUA in 1949 was not at all clear. When Bohm's friends advised him to "take the Fifth," he objected that the American Communist party was not an illegal organization, so there was no question of self-incrimination in admitting that he had been a Communist. On the other hand, he could not refuse to give evidence, or choose to answer some questions and not others, without facing a charge of contempt of Congress.

Lomanitz and Bohm met in Princeton to discuss what could happen to them when they appeared before the committee. While they were walking along Nassau Street, they bumped into Oppenheimer, who was just leaving a barber shop. When they discussed their dilemma, Oppenheimer said, "Oh, my God, all is lost.[4] There is an FBI man on the Un-American Activities committee." Lomanitz found his former supervisor to be "paranoiac" and was troubled as to why he appeared so disturbed. Oppenheimer's own account of the incident, however, was subtly different: testifying before the Personnel Security Board of the Atomic Energy Commission, he insisted that he had told his former students to tell the truth, to which they had replied, "We won't lie."[5] Of course, by this time Oppenheimer's opponents were attempting to build a case against him, asserting that he was actively supporting people like Bohm and Lomanitz.

After two postponements, Bohm was finally interrogated on May 25, 1949, in an executive session of hearings on "Atomic Espionage at the Berkeley Radiation Lab." Bohm's counsel was Clifford Durr, who, as Bohm recalled, had been recommended by E. U. Condon, director of the National Bureau of Standards. Over the next months Condon, who had earlier defied the committee, referred many young scientists to Durr as they became involved with the HCUA. Condon gave important support to Bohm and Lomanitz during this period.

Durr was noted for his work in civil rights, including the defense of those accused of un-American activities. Many of these young people, who considered themselves to be loyal Americans, experienced considerable trauma at appearing before the committee and even facing arrest and incarceration. Some, he discovered, had nervous breakdowns, and Durr helped them find psychiatric help. As Bohm recalled, if Durr asked for a fee, it was very little—just sufficient to cover his expenses.

Bohm first met Durr at Condon's house in Washington and found him sympathetic. Durr now counseled Bohm that if he took a position like Frank Oppenheimer's—answering questions about himself but refusing to answer questions about others—he would probably go to jail. His only possibility was to take the Fifth Amendment.

Accompanied by Durr, Bohm faced a committee chaired by John S. Wood of Georgia and which included Richard M. Nixon. Many years later, when Nixon was forced to resign the presidency, Bohm was uncharacteristically gleeful, saying that the last member of his investigating committee had now been accounted for.

The committee's business was to determine just what had been going on at the Radiation Laboratory; had there indeed been a Communist cell, and had a "secret atomic formula" been passed to the Russians? Secret formulae fired the public imagination during the cold war—they were the stuff of Dick Tracy comics, radio serials, B movies, and the yellow press. To the physics community, however, the notion that passing a single formula could give away a complicated technical secret was laughable. Building an atomic weapon required hard work and overcoming considerable engineering difficulties. The actual principles involved—the "secret formula," if you like—were widely known to scientists throughout the world and were not particularly secret at all. The idea that one could write down on a single piece of paper sufficient information to enable the Russians to build a bomb of their own was absurd.

The committee members, however, were neither scientists nor engineers. They had the bit between their teeth, and they were going to ferret out Communists and traitors. In retrospect, possibly, one should not be too hard on them. If they genuinely believed that the nation's most vital secrets could be written down on a single sheet of paper, and that scientists working in a key installation had been willing to pass on such a piece of paper, then they must have felt justified in their duty to uncover the truth.

Faced with the committee, Bohm gave his name, educational background, and the fact that he had been employed by the Manhattan Engineering District from the fall of 1942 to the fall of 1946. When asked if he knew any persons on a list of names, Bohm replied in each case, "I do not recall any such person."[6]

Senior Investigator Louis J. Russel then asked, "Mr. Bohm, have you ever been a member of the Young Communist League?"

BOHM: I can't answer that question on the ground it might tend to incriminate and degrade me, and also, I think it infringes on my rights as guaranteed by the First Amendment.

CHAIRMAN JOHN S. WOOD: What particular rights guaranteed by the First Amendment?

MR. DURR: He is not a lawyer. I think it would be freedom of assembly and association and freedom of speech.

The committee them pressed Bohm on whether he knew Steve Nelson.

BOHM: I want to answer it in the same way, by saying I agree fully with my counsel; I back up the statement of my counsel. I can't answer the question on the ground it might tend to incriminate and degrade me.

Another question on his politics; Had he ever been affiliated with any political party or association? To their surprise, Bohm answered yes, admitting, "I would say definitely that I voted the Democratic ticket."

After unsuccessfully pressing him on other names, the committee excused Bohm from the hearing room.[7]

Two days later Princeton University issued a statement in his support. Bohm's work was "zealous and effective," and he had "earned the respect of his colleagues"; he was regarded "by his Princeton colleagues as a thorough American and at no time has there been any reason for questioning his loyalty."[8]

On June 10 Bohm appeared for a second time, and now he sounded a little tetchy. When asked about his personal and professional past, he replied, "I believe you have all this. . . . Well, you already have the résumé. Would you please go through it, and I could correct it if necessary."

Again he was asked, without success, about membership and certain names. When he was about to be excused, Bohm addressed the committee: "I would like to make a statement concerning my loyalty—that I have been completely loyal to the United States during all my life and that I would not contemplate any disloyal acts, and I know of no disloyal acts. If I knew of any, I certainly would not countenance them."

This statement gave the committee new grounds for attack. Since they believed that classified information had been leaked from the Radiation Laboratory, their duty was to uncover the facts in the case. As Chairman Wood pointed out, Bohm had protested his loyalty yet at the same time refused to answer the committee's questions: "it is just a little incongruous that a person can take these two positions and still say he is loyal to the American people and the American government."

BOHM: It doesn't seem incongruous to me. I have stated that I know of no effort to obtain information; that if I knew of them I would tell of them immediately. Now, this is an entirely different question, it seems to me.

Such a protestation, Bohm was told, carried no conviction unless he was willing to aid the committee.

BOHM: Yes, I believe it does. That is, I mean I think that essentially—I feel on the basis of what I know that I do not believe that these other questions would actually be of any help, but I am not sure about that. I mean I am not an expert, but—

He was then asked whether, if he were convinced that his answers would help the committee, he would give information. Bohm answered that he would have to think about that in each specific case. Francis E. Walter argued that the committee itself was the best judge of the course of questioning and that Bohm should help them "protect our country."

BOHM: I am certainly not an expert on security measures, but different people have different opinions as to what are the best security measures, and many people I know have felt that security measures are quite adequate, and some even feel that they have been so over emphasized as to interfere actually with the security of the Nation. This is a question upon which very many different people have different points of view.

MR. MCSWEENEY: You mean overanxious in making things secure?

BOHM: I believe that in some cases many people feel that security has been so—people have concentrated so much on security that they are not able to do the job on hand. In other words, I mean, as an analogy, take the individual who was so afraid to cross the street he would never be able to do anything. You would have to take a certain medium attitude in that.

MR. WOOD: In other words, the question of the security of the

American Government and its institutions and its restricted information has been overemphasized in your opinion? Is that right?

BOHM: I wouldn't care to make a categorical statement, because I have not studied it adequately. I would say it probably has in some cases and in other cases probably not.

MR. MCSWEENEY: Do you not think that persons who have to classify information had better err on the side of overcaution than the other side?

BOHM: To a certain extent, but there is always a limit. You have got to draw a dividing line somewhere.

After a little more debate on the issue, McSweeney conveniently dug himself a hole, into which he fell.

MR. MCSWEENEY: I see how it could be carried to a conclusion where it would interfere with efficiency, but, on the other hand, I say it is better to err on the right side.

BOHM: I say it is better not to err at all.

Bohm's investigation was adjourned until further notice. Following the adjournment, Bohm believed that the issue would die away and that he could go back to his research.

Bohm's views on bureaucrats and government institutions, however, were becoming particularly cynical. He attended a lecture on mathematical paradoxes, in which the speaker referred to a famous paradox of Bertrand Russell on classes of objects, and on classes of these classes. (Stated in popular form, Russell had asked, "If the barber is the man who shaves all men who do not shave themselves, then who shaves the barber?") Afterward Bohm joked that Congress should appoint a committee to investigate all committees that did not investigate themselves. This, of course, left open the burning question of whether that committee should also investigate itself!

Despite Bohm's optimism about being able to return to his

research, political events were rapidly making that unlikely. On September 23, 1949, President Truman announced that the Soviet Union had tested an atomic device. On September 27, HCUA counsel Frank Tavenner interrogated Irving David Fox, one of Bohm's former friends at Berkeley, and said that the hearings into the Radiation Laboratory had now become of extreme importance. Two days later the committee released its "Report on Soviet Espionage,"[9] which charged that Joe Weinberg had given Steve Nelson "atomic secrets," and that Nelson in turn had sold them to an employee of the Soviet embassy. Bohm, Lomanitz, Fox, and Weinberg were all named as members of a cell that met with Steve Nelson in August 1943. The report recommended that Weinberg be indicted for perjury.

At the same time the United States was about to go to war with Korea, and anti-Communist feeling was mounting. Alger Hiss was convicted of spying in January 1950, and Julius and Ethel Rosenberg were arrested later that year. Fifty-six persons were cited for contempt of Congress by the HCUA, and on August 10 Chairman Wood demanded that Bohm and others be cited as well. Bohm and Lomanitz were both indicted by a federal grand jury, and on the evening of December 4, 1949, David Bohm was arrested. The marshal was friendly, and when Bohm asked him for advice, he was told that they could drive to Trenton, the capital, to try to obtain bail. When Bohm inquired of the marshal how high the bail would be, he replied that it would be $20,000. This seemed excessive, since Bohm knew that other people had been given bail of only around $1,000. He wondered what the authorities had against him. In any event the marshal allowed Bohm to call his lawyer, who said that he would look into the bail question.

Back en route to Trenton, the marshal talked about science, asking his prisoner about Einstein. He had come from Hungary, he said, and was a loyal American; he hoped that Bohm had not been disloyal. In Trenton they waited at the courthouse until Bohm was arraigned before U.S. Commissioner Walter B. Petry and charged on eight counts to appear in federal district court on December 8. Hearing what had happened to his friend, Sam Schweber drove over to Trenton to help. Finally the U.S. attorney in Trenton received a call from Washington saying that bail could be set at $1,500. Bohm made out the check—Schweber recalled that he lent Bohm the money for the bail.[10]

As Bohm was returning to Princeton, the university administration was meeting behind closed doors. News of Bohm's indictment had not even reached the newspapers before Princeton's president, Harold W. Dodds, released a statement suspending the physicist "from all teaching and other duties" for the duration of the trial. An ominous footnote mentioned that Bohm's appointment was due to terminate in June 1951. On the morning after his return from Trenton, Bohm received a letter forbidding him to set foot on the campus or to lecture to the students.

Rossi Lomanitz, for his part, was fired from his position at Fisk University. He went to Oklahoma City, where he took a job as a laborer at a local power plant. When reporters were tipped off about his place of work, he was forced to leave. From that time on Lomanitz could obtain only laboring jobs. Whenever something better appeared in the offing, the potential employer would receive a visit from federal agents.[11]

Why had Princeton now turned against Bohm, when previously they had issued a statement of support? Private universities like Princeton depended heavily upon benefactors and corporate philanthropists. As far back as the Red Scare of 1919–1920, universities had been warned against employing Communists, and a number of faculty had been demoted, dismissed, or "resigned." In 1950 institutions like Princeton that were receiving government research contracts were eager to prove their political purity.[12]

In Harold W. Dodds the university had a president who was willing to toe the line. Speaking at the University of Hawaii in 1949, he argued that Communists had surrendered "their rights as persons, made in the image of God, and submit[ted] to such slavery of bodies, minds and souls." Among communists, "loyalty to one's country is made a despicable thing," he continued. "Treason is the accepted code of conduct of all practicing communists."[13] Speaking the same year in San Francisco, he claimed that Communists were unfit to teach in schools or universities, for they were "part of an international conspiracy. The communist doctrine denies academic freedom. Its followers cannot be honest." Dodds, then, was hardly likely to be sympathetic to David Bohm, who, in addition to his left-wing background, had another strike against him. Princeton's administration was not without its anti-Semitic elements. When Silvan S. Schweber tried to meet with Dodds, the president simply "showed him the door."[14]

Bohm himself continued to meet with his students, despite the ban, and wandered onto the campus to visit the library. Exile was not totally negative, for he now had a great deal of spare time in which to pursue his research. Even though he was suspended from teaching, Bohm believed that he had the active support of his colleagues and students. Allen G. Shelstone, the physics department chairman, assured him that he would continue to receive his full salary, and a week after his suspension, thirty-eight of the forty-seven graduate students in the department signed a letter supporting Bohm. This letter was printed in *The New York Times* (December 16), *The New York Herald Tribune* (December 18), and *The Daily Princetonian* (December 15).

The charges against Bohm and Lomanitz were for contempt of court (by a grand jury that had been empaneled on October 3, 1950, and sworn in on October 4, 1950). Their offense was the refusal to answer questions concerning, among other things, Joe Weinberg's alleged membership in the Communist party and his association with Steve Nelson. Since at the time of Bohm's indictment the courts had not yet tested the validity of an appeal under the Fifth Amendment, Bohm faced the possibility of a jail sentence. While he was engrossed in physics, Bohm appeared calm and focused. Yet he was experiencing considerable anxiety and fear, going over and over the possible outcomes of his trial and even wondering if he should flee the country. His friends saw his anguish and tried to help him as best they could. Schweber, for example, contacted the Israeli embassy to find out whether the U.S. government could extradite Bohm should he seek asylum in Israel.

In search of a lawyer, Bohm visited the American Civil Liberties Union, but his case was turned down.[15] Ultimately he was represented by one of Clifford Durr's assistants. On December 11 the Supreme Court decided, in the case of *Blau v. the United States,* that the right against self-incrimination had existed before congressional committees and that impelling a petitioner to testify ran counter to the Fifth Amendment. Based on this decision, Bohm's lawyer moved that the court dismiss the charges against him, but the bid was unsuccessful. His trial, originally set for January 18, 1951, was rescheduled. In the meantime James J. Matles, of United Electrical Workers Union, was acquitted of charges similar to Bohm's by a U.S. district court.

On May 30, the day before his trial began, Bohm traveled to Washington and stayed with Ed Condon. By now he was worried that the pressure being exerted upon him to testify was connected with the renewed security interest in Oppenheimer. Were the authorities trying to build a case against the distinguished physicist—with Bohm as a potential witness? On May 31 Bohm sat in the courthouse waiting nervously while other cases were dealt with. During that time his lawyer talked with the prosecuting attorney, even discussing some of Bohm's scientific ideas. Finally the case was called. In the witness box Bohm felt that the prosecutor was trying to trap him by going over his evidence before the HCUA. In the end, however, he was acquitted on all counts.

His friend Rossi Lomanitz was also acquitted. He recalled that at his own trial, the judge lashed out at him with particular vigor. Fearing it meant a guilty verdict, Lomanitz had been about to protest when his lawyer held him back, whispering, "This means he's going to let you off."[16] Joe Weinberg was brought to trial in 1953, but the HCUA refused to hand over evidence to the court, and he too was acquitted.

Even though Bohm had been acquitted, Princeton University made no announcement that his teaching duties had been reinstated, and when his contract came up for renewal, in June, it was terminated. In those days young faculty members were generally placed on contract for several years, and the contracts were renewed automatically provided that their teaching duties and research were satisfactory. Bohm had already begun to distinguish himself as a physicist of considerable talent; he was supervising a number of graduate students, giving lectures, and was in the final stages of completing a textbook. Renewal of his contract should have been a foregone conclusion. Clearly the university's decision was made on political and not on academic grounds.

In coming to his decision, President Dodds had been placed under considerable pressure. Millionaire Robert Wood Johnson, of Johnson & Johnson, wrote to the university saying that "an institution seeking private support should at this time put its house in order." Letters to the *Princeton Alumni Weekly* argued that the university must preserve its image; "We want neither criminals nor cowards as teachers of our youths," read one. Others, however, took the opposing side, arguing that it was necessary to preserve academic freedom.[17]

In 1962, when Dodds was interviewed for the Columbia Oral History Project, he attempted to exonerate himself by claiming that Bohm's contract had gone unrenewed for professional rather than political reasons: "In his own scholarly life, he [Bohm] didn't merit promotion"—a particular irony in the light of Bohm's distinguished career.

Bohm's colleagues may not have been as vigorous in supporting him as Bohm believed at the time. While a majority of the faculty did vote to reappoint him, a number of them, later sitting as the Committee on Appointments and Advancements, overturned this decision. The physics department appears to have been less interested in Bohm's fate than in the fact that it had not been included, from the beginning, in the administration's decision. One department member, Martin Schwartzschild, later claimed that "a sham group of leading professors" had been formed that "got the administration to agree that no such dismissal should occur in the future without involvement of the appropriate faculty committee."[18]

While the department may have had some complicity in Bohm's removal, at least according to the researches of Russel Olwell, in the last analysis it did not have that much power. President Dodds had already consulted Henry Smythe, a former chairman of the physics department and the man who, Roy Britten remembered, boasted that he owned two dogs who would bark whenever anyone mentioned the word *Communist*. Smythe advised Dodds that there were insufficient grounds for dismissing Bohm. But Dodds replied that while he was in general agreement with Smythe, there was a point of exception: "I am unwilling to agree that nothing has happened to date to bear on the reappointment of the man in question. . . . This news I have conveyed to Allen [Shelstone]," the department chairman.[19] Dodds had made his position clear to the department, even before it began its own deliberations. The faculty were therefore free to vote as they wished, since their decisions would be merely symbolic. In Kenneth Ford's opinion, it was Bohm's fate that led some physicists to form the Scientists' Committee on Loyalty.

CHAPTER 7
Hidden Variables

HAVING LOST HIS UNIVERSITY POSITION, Bohm now looked around for another. Henry Smythe, the former head of the physics department, attempted without success to obtain a position for him in one of the southern universities. Scientists at Bell Laboratories and RCA tried to hire him, but senior management who feared the loss of government contracts turned him down. Lyman Spitzer wanted Bohm to join Project Matterhorn, which sought a way to control nuclear fusion using high temperature plasmas—Bohm was the world's leading theoretician in this field. Had he been hired, David Pines later estimated, "the U.S. effort might easily have been saved some tens of years and millions of dollars."[1] But again a politically tainted physicist was someone to be avoided.

Einstein was also interested in having Bohm work as his assistant at the Institute for Advanced Study, a body distinct from Princeton University. It was Einstein who had said, referring to the need for a radical new quantum theory, "if anyone can do it, then it will be Bohm."[2] Oppenheimer, however, overruled Einstein on the grounds that Bohm's appointment would embarrass him as director of the institute.[3]

Einstein then wrote to P.M.S. Blackett at the University of Manchester, England, recommending Bohm for a position. Bohm has "a clear mind," he wrote, "is very energetic in his scientific work and of a rare independence in his scientific judgement." A second reason to give Bohm the chance to work outside the United States, Einstein continued, was that he had refused to answer official questions concerning his colleagues: "This admirable attitude was the cause of an official indictment and subsequent termination—or rather nonrenewal—of his appointment at Princeton University."[4] But Einstein's effort on Bohm's behalf was without success.

Mort Weiss recalled that "around the time Bohm was having problems with the committees," he made a trip to Florida. Bohm had been staying for a few days with his friend on Long Island, and Weiss had the feeling that his friend was hiding out from the FBI. One evening Bohm asked Weiss to look out the window and see if a yellow convertible was driving back and forth. Weiss saw the car. "Are you being pursued?" he asked.

"Yes," Bohm replied, "they are looking for me."

He then said he had to go to Florida, where someone would help him get out of the country, as far as Weiss could recall; he may even have said that this person would help him get to Brazil. The two men waited until it was dark and walked to the subway at around eleven o'clock. On the train Mort Weiss noticed a man reading the back page of a newspaper. On the front of the paper was a photograph of Bohm, with words to the effect that "all they ever got from him was his name." Weiss joked that at last Bohm had gotten his name in the papers, but his friend was far from happy. When they arrived at Penn Station in New York City, Bohm bought a ticket for Florida, and that was the last Mort Weiss saw of Bohm for many decades. He assumed that his friend had traveled on directly from Florida to Brazil.[5]

In Florida Bohm worked briefly for a small industrial laboratory. The people he met were "wealthy but quite friendly. They want me to stay, but I just couldn't put my heart into their kind of life. It just frustrates something very fundamental in me." But at least he could relax by the ocean: "I like to float and watch the waves break. One gets a feeling of unity with this warm sea, and sometimes I wish I could dissolve in it and spread out to its furthermost shores."[6]

Those last months at Princeton, from his arrest to his trial and final suspension, were a difficult period for Bohm. Only while doing physics could he lose himself and, for a time, forget his fear. Even when he had been correcting the proofs of *Quantum Theory,* it had been "hard to concern myself with getting all these formulas correct."[7]

The book itself was published during Bohm's final months in Princeton, and in its preface Bohm acknowledged his debt to Oppenheimer. His desire, he wrote, was to "express the results of the quantum theory in terms of comparatively qualitative and imaginative concepts"—although some of these concepts admittedly would be radically different from anything previously held in physics. The book also introduced a theme that was to run through all of Bohm's later work, that of wholeness: "quantum concepts imply that the world acts more like a single indivisible unit, in which even the 'intrinsic' nature of each part (wave or particle) depends to some degree on its relationship to its surroundings."[8]

This notion of holism, which is something of a cliché today, was far from being widely accepted in 1951. The word itself had been coined in the early years of the century by the (selectively racist) South African politician Jan Smuts, but it did not really come into prominence until the 1960s, in the hands of writers such as Arthur Koestler.

The building blocks in a Newtonian world are particles—distinct individual entities—in interaction. It is a world of objects and independent elements of reality. By contrast, the quantum world is fundamentally holistic, something that is inherent in the very notion of the quantum itself. The quantum is indivisible, and in every observation of the quantum world, the observer and the observed are linked by a quantum. Since the quantum cannot be broken apart in any way, during an observation the observer and the observed must be considered an undivided whole. In the quantum domain independent entities are not objects but secondary things that emerge out of the theory at some limit. They exist only as approximations arising out of the whole.

As a boy, Bohm had watched the vortex of water spin down the drain of his bathtub. This vortex appeared to be a stable and localized entity, yet did not exist apart from the bath water. Likewise, while working at Berkeley, he imagined the electron as a wave that collapses inward from the entire universe. In his vision of the cosmos, each event, each appearance within the world, was a manifestation of some much larger process. He could even enfold that whole within his own body and through subtle muscular dispositions sense the laws that governed the universe. Now, in his textbook, Bohm was writing of a wholeness that extends all the way down into the microworld. This vision of wholeness was to become the leitmotif of all Bohm's research.

Most textbooks on quantum theory begin with the basic equations of the theory and move rapidly into problem-solving and application. It is a mark of Bohm's unique approach that the fundamental equation of the theory, Schrödinger's equation, does not appear until page 191. For almost two hundred pages Bohm takes the reader step-by-step through the sorts of experiments and arguments that had first convinced physicists that classical physics must be replaced by something radically new. Where mathematics does appear, it is the kind that is accessible to any undergraduate in a first-year physics course. The desire to understand the meaning of phenomena always drives the theory, Bohm shows; and the theory never leads the physicist.

In his last two chapters, Bohm turns to the difficulties of the theory of measurement. In particular, he shows that quantum results must always be described in the language of classical physics. "We conclude then that quantum theory presupposes the classical level and the general correctness of classical concepts in describing this level; it does not deduce classical concepts as limiting cases of quantum concepts."

This very important point, which had earlier been made by Bohr, deserves some explanation. Einstein, in his special theory of relativity, realized that classical Newtonian physics breaks down at speeds close to that of light and in strong gravitational fields. Clearly a more general theory (relativity) was required, which would explain phenomena at high speeds as well as contain the older classical mechanics as a limiting case for slower speeds and weak gravitational fields. In other words, classical mechanics is contained within and can be derived from Einstein's relativity.

Physicists had also discovered that classical mechanics breaks down in the region of the very small and that a new (quantum) theory was required. One would have hoped, by analogy, that classical physics would also be contained within quantum theory, emerging as a limiting case for bodies much larger than atoms. But at this point things became complicated. All quantum observations and measurements are, of necessity, made in laboratories using dials, meters, computer print-outs, and the like. In other words, quantum observations demand the use of laboratory apparatus that is well defined in terms of classical physics.

In this context, Niels Bohr also observed that scientists do not communicate exclusively through mathematics but also use ordinary, everyday language. Physics is a social activity in which scientists discuss the meaning of their theories and the observations and measurements they make. Even when talking about the quantum world, such discussions can take place only in ordinary language, albeit refined by concepts coming from classical physics. Yet paradoxically, the inner structure of laboratory apparatus (classical objects) used to measure quantum systems can be described in quantum terms. Thus, while physicists would like to have classical mechanics be a special limiting case of quantum mechanics, at the same time every quantum parameter has to be defined with respect to large-scale classical apparatus. Bohm gave considerable thought to these paradoxical questions and, as he later put it, attempted to put Bohr's interpretation in the best light. (It was only in the 1960s that Bohm began to realize the book had leaned closer to Pauli than to Bohr.)

In these final chapters Bohm also recast one of the major objections that had been made to quantum theory by Einstein and his colleagues Nathan Rosen and Boris Podolsky. Einstein believed that physics must deal with a reality that is totally independent of the human observer. In this context he argued that systems in distinct regions of space, far from the influence of other systems, must each have a well-defined and independent reality. To this end, he proposed a hypothetical experiment in which a pair of quantum particles are separated by a large distance where, he assumed in contradiction to quantum theory, common sense dictates that each must exhibit independent properties. Unfortunately, the EPR paper (Einstein, Podolsky, Rosen) does not present the paradox in the best possible way. Bohm therefore recast the paradox in a clearer and more provocative form.*

Einstein and his colleagues had presented the paradox in terms of

* When the paper was composed, Einstein had not yet learned sufficient English, so his ideas were set down by his collaborator, Podolsky, who also was not fluent in the English language.

measurements of the position and momentum of a pair of correlated particles. (Technically speaking, these are continuous variables, or eigenvalues, which makes it difficult to understand the full force of the paradox.) Bohm's leap was to recast the argument in terms of spin, which can have only one of two possible values. Immediately the full force of the paradox became apparent. Later John Bell, after reading Bohm's papers on hidden variables, presented an additional reformulation of the EPR paradox that could be tested in the laboratory.

Upon publication, *Quantum Theory* gained favorable reviews and was adopted by a number of universities for their courses. Bohm sent copies to several leading physicists. Pauli replied with a warm letter approving of the way Bohm had woven mathematics and physics together and addressed philosophical questions. Bohr did not write back, and in retrospect Bohm wondered if that was because the book's approach had been too close to that of Pauli and had not accurately reflected the Danish physicist's own approach.

The most important reaction came from Einstein, who told Bohm that he had enjoyed the book, feeling that it presented the Copenhagen interpretation in the clearest way possible. Nevertheless, he still could not accept the theory and asked Bohm to visit him. As Einstein presented his objections, Bohm's own misgivings about the theory crystallized.[9]

Einstein was the elder statesman of physics, but among Bohm's contemporaries he had come to be regarded as an anachronism, a remnant of the classical age of physics that had been forever swept away by quantum theory. As a young man, Einstein had overthrown Newtonian physics to replace it with his theory of relativity. Now he had become the epitome of the old guard, attempting to hang on to determinism and an objective reality. Bohm, for his part, had great sympathy for Einstein's position.

A few years earlier (in 1946), in a brief autobiographical essay,[10] Einstein had acknowledged that when it came to predictions, quantum theory was "the most successful physical theory of our period." Nevertheless, it was unable to give a picture of "physical reality" that was independent of any human observer. The fundamental descriptive element in the theory is the wave function, but it was not at all clear to

Einstein how this element could ever correspond to "a real factual situation." When a measurement of some property is made in our everyday world, we naturally assume the system actually "has" the property in question. Not so in quantum theory, which is incapable of answering questions about the reality of descriptions. For these reasons Einstein felt that the theory must be incomplete.

It is difficult to understand Bohm's motivations and thinking at the time he spoke to Einstein. In his book he presented quantum theory in a particularly convincing way. Yet when Bohm arrived at Berkeley from Caltech, he had been "a confirmed classicist," unconvinced by quantum theory, as Joe Weinberg recalled. From the perspective of the 1970s, looking back on his Princeton days, Bohm himself referred to doubts he had had about the theory. Yet the essential holism of the quantum theory must have been particularly appealing to him, as well as its rejection of mechanistic ways of thinking. He even joked that the theory should be called "quantum nonmechanics."* His book does appear to be positive about quantum theory and even goes as far as to explore analogies between it and thought processes.

Why then, after talking with Einstein, did he reject quantum theory and search for an alternative? One reason may have been Bohr's denial of the possibility of explaining the underlying actuality of quantum events. Another may be the theory's rejection of causality. (Again it is important not to confuse Bohm's desire to retain causality with his rejection of mechanistic theories.)

It is also true that Bohm's political views did influence his physics. He was clearly attracted to the wholeness of quantum theory and to the subtlety of Bohr's thinking. But from his Marxist perspective, the theory left much to be desired. For Bohm, science could never be a purely abstract study; it had to have practical and social importance. Physics enables people to understand the nature of matter, in his view, and a proper understanding of matter is essential to an ordered Marxist society. From this perspective Bohr's Copenhagen interpretation

* Since Bohm's position has frequently been misunderstood, it is important to emphasize that he always rejected mechanistic thinking and, in the early 1990s, was horrified when his own hidden variable theory was later called "Bohmian mechanics" by some physicists.

appeared overly metaphysical and even quasi-mystical. Rather than presenting the subatomic world in a clear way, it seemed to dissolve matter into the insubstantial. Bohm was also concerned about the implications of denying causality in favor of absolute chance; Marxist theory too is deeply based on causal effects on the individual and society.

For whatever motivation, after Bohm talked to Einstein, he began to wonder if there was a way to explain quantum processes without the mystifying assumptions of the current theory. A number of other physicists had also been dissatisfied with the quantum theory, but none had been able to create a viable alternative, despite talk of the possibility of "hidden variables." Einstein himself had devised such a theory but had withdrawn it before publication.

The idea of hidden variables can be understood by analogy to Brownian motion. The nineteenth-century Scottish botanist Robert Brown (1773–1858), while looking through a microscope, had noticed the ceaseless agitation of tiny grains of pollen suspended in water. These movements appeared to be completely random.* A variety of explanations were advanced at the time including, for example, the notion that they were evidence of a mysterious life-force. It was left to Albert Einstein, carrying out his first theoretical research project, to discover the correct solution.

The Austrian physicist Ludwig Boltzmann had already proposed the existence of submicroscopic molecules as an explanation for the behavior of gases. The physicist and philosopher Ernst Mach objected that these molecules had no actual material existence and were merely theoretical postulates used to derive the properties of gases. Einstein took the molecules literally and asked what would happen if a tiny pollen grain were constantly bombarded by water molecules. The result, he demonstrated, would be tiny, random jumps of the pollen grain. While molecules are hidden from us because of their size, it is possible to infer their existence by the visible effects they produce on the random movement of tiny pollen grains.

* One sometimes notices dust motes dancing in a beam of sunlight. This is not true Brownian motion but is the result of dust being seized by tiny currents of air circulating in any room. The observation of Brownian motion requires carefully controlled conditions.

Could it be that the probabilistic results of quantum theory were the result of the underlying (deterministic) motion of yet smaller particles—hidden variables? Bohm soon realized that a coherent theory could not be won in such an obvious way. Quantum theory involved radically new results that could not be obtained by any simple or obvious extension of classical theory to smaller domains. If he were going to create a new theory of hidden variables, it had to be subtler than a subquantum level of deterministic particles.

Quantum theory had revealed that sometimes the electron behaves like a wave and sometimes like a particle. Bohm, well steeped in classical physics, knew that something analogous to this wave-particle theory already existed. In the nineteenth century the Irish mathematician W. R. Hamilton had shown that it is mathematically possible to recast Newton's laws, about the movement of particles, into a description involving waves. Using mathematical transformations, he could move back and forth between a particlelike and a wavelike description, yet always retaining the same results for calculations.

Hamilton's approach was used in quantum theory, Bohm also knew, as a certain approximation to simplify and speed up calculations. The WKB approximation, as it was called (after G. Wentzl, H. A. Kramers, and L. Brillouin) occupies a position midway between classical and quantum mechanics. By making the assumption that quantum particles move along actual trajectories, it offers a simple way of calculating these paths.

All physicists agreed that such a picture—electrons moving along trajectories—had no objective validity, being no more than a mathematical trick used to calculate an approximate result. Bohm, however, wondered what would happen if the approach were taken literally. What if electrons really did move along paths in a totally deterministic way? What would have to be added to Hamilton's approach in order to transform this approximation into something deeper that would reproduce all the results of quantum theory exactly?

Bohm went back to the fundamental equation of quantum theory, Schrödinger's equation, and transformed it so that it would produce the sorts of trajectories similar to those found in the WKB approximation—but without making any approximations. He did this by introducing a radically different sort of potential (or field) into physics, which he called "the quantum potential." This quantum potential, he realized, is responsible for all the bizarre results we normally associate with the quantum domain. No longer, Bohm argued, was it necessary to invoke metaphysical arguments about Heisenbergian potentialities and actualities, collapsing wave functions, and irreducible probabilities. Quantum nature could now be explained in an entirely causal way. Particles could be shown to move along well-defined paths, provided that the existence of a new sort of potential were admitted, one with curious holistic properties that act to guide electrons. And what of the famous uncertainty principle, and the probabilistic results of the theory? They followed from the enormous complexity of the quantum potential itself, which acted to randomize the electron's motion.

Bohm was enormously excited by what he felt was a major discovery. "I can't believe that I should have been the one to see this," he told Miriam Yevick.[11] His new theory would transform the way everyone thought about elementary matter. By eliminating what he now felt to be Bohr's confused and mystifying Copenhagen interpretation and replacing it with a clear and rational explanation for the nature of matter, he would also satisfy the Marxist program. Once people understood that causality operates right down at the level of the atom (albeit in a nonmechanistic way), he believed, they would be impelled to order human society in a more rational way.

Bohm worked quickly during 1951 to develop the theory of his hidden variable approach. To show how the theory worked, he applied it to the hydrogen atom and wrote a preliminary version, which he sent to Bohr, Pauli, Einstein, and several other physicists. From Louis de Broglie in Paris, Bohm received a somewhat surprising reply.[12] It turned out that de Broglie himself had once proposed a similar idea at the 1927 Solvay conference but that Pauli had made such strong objections that the French physicist had abandoned the work. In his letter to Bohm, de Broglie outlined the objections that Pauli had made to the theory.

The next reply was from Pauli himself, who repeated his objections in detail.[13] They mainly involved what would happen when the theory was generalized to more than one electron, but by no means did they deter Bohm, who went on to develop his theory for situations in which many particles were in interaction. Confident that he could explain the actuality of quantum measurement without the need for a bizarre "collapse of the wave function," he mailed two papers to the *Physical Review,* where they were scheduled to appear in Volume 85 in 1952.[14]

These papers demonstrate the basic theory and show how it could be applied in a few key areas. Bohm continued to work on the theory later, when he was living in Brazil and Israel, but he did not develop its full implications until the late 1970s and 1980s, at Birkbeck College. After his death research groups in several countries focused on aspects of his approach—one even called it "Bohmian mechanics," somewhat of an irony in view of Bohm's own belief that quantum phenomena are inherently nonmechanistic.

Why did it take Bohm so long to work out all the implications of his theory? Here a world of explanation is necessary, for although it may seem unusual to the nonscientist, the creator of a new theory may not necessarily fully understand what he or she had done.

A scientific theory may be compared to a deeply satisfying poem. Such a poem can be visited again and again during one's life. At each reading its density, allusions, images, and resonances are unfolded. The best poetry is an inexhaustible inscape, and its metaphors suggest a variety of different readings, none of which brings closure to its range of potential meanings. In similar fashion a scientific theory exists partly as mathematical formulae and partly as a cloud of words surrounding the formulae and their variables. Like the poem, the theory's meaning, or the way it functions, can be unfolded in a variety of different ways. Wittgenstein said that the meaning of a word lies in its function, in the way it is used within language. To some extent the meaning of a scientific theory depends on how it is used. The task facing Bohm was to unfold the theory and the implications of its formulae in a variety of different ways, none of which were apparent at the time of its creation.

That a creative mind may not necessarily be aware of what it has given birth to is as true in science as in the arts. Speaking in 1970, after having been awarded the German Shakespeare Prize, the playwright Harold Pinter attempted to explain how characters in his plays take on their own life and, in a sense, observe their writer. Thus, he said, the reason a character makes a particular remark may not be at all clear to him (the author) and will emerge only during the rehearsal.

Over the next forty years, Bohm pushed his hidden variable theory in new directions. At first he presented it as a defense of determinism, but later it took on ever more subtle forms and ended up being an intimation about the activity of information in the universe. Bohm himself expressed these changes in the differing names he gave the theory; hidden variable, the causal interpretation, and finally the ontological interpretation. Yet at the same time, the theory existed independent of Bohm. At times it could resist him, yet as he probed into its functioning, it gradually revealed more of its inner meaning.

Bohm's first version of his theory did not officially appear in print until after he left the United States. In mathematical terms the papers explained the basis of his theory and demonstrated how it reproduces all the predictions of conventional quantum theory. They left open the door for new experiments that might one day decide between his own theory and conventional quantum theory. Conventional quantum theory has been tested only down to distances as small as 10^{-13} centimeters, and Bohm suggested that at smaller distances the two theories might part company. One possibility for differences between the two theories involved the nature of quantum probabilities. Quantum theory gives no explanation for these probabilities, simply assuming that they are absolute. Therefore two measurements, no matter how soon the second one is taken after the first, show no correlation between them. By contrast, Bohm's theory claims that probabilities are a measure of our ignorance of the complicated and sensitive motion of electrons. Following one measurement, the system rapidly randomizes, with the result that successive measurements appear uncorrelated with it. But if two measurements are carried out within a sufficiently short time of each other, there will not be enough time for this total randomizing to take place, and a slight correlation should be observed between successive observations. As experimental techniques were refined, Bohm hoped that one day a test would be conducted to decide between his and the conventional theory on this point.

Bohm's approach had also made explicit what had hitherto remained only implicit within quantum theory: the notion of wholeness. Bohr had, of course, written about wholeness, but now Bohm was able to show that it must follow inevitably from the nature of his quantum potential. Forces and potentials in nature fall off rapidly with distance. The closer a nail approaches a magnet, the stronger is the magnetic field it experiences and the greater the pull. But the effects of Bohm's quantum potential do not fall off with distance. The effect of the field depends not upon its strength but upon its "form," which means that quite distant objects can still affect each other. In other words, even distant parts of quantum systems are intimately linked through the quantum potential. The question then becomes not so much why things behave as wholes as how independent objects exist at all. In this period Bohm did not fully appreciate the revolutionary nature of this aspect of his theory; only later was "nonlocality" proved to be an essential aspect of the quantum potential.

Bohm awaited the publication of his theory, anticipating that the physics community would react with enthusiasm. At the same time he was still having difficulties with his immediate career. A professional position seemed impossible in the United States, and his inquiries in Europe had drawn a blank. Two Princeton graduate students, Tod Staver and J. Tiomno, were working at the university in São Paulo, and Bohm now wrote to them asking about possibilities in Brazil. Their reply was encouraging, and in August 1951 Bohm asked Einstein to send a letter of recommendation to Professor Abrahão de Moraes in the department of physics at São Paulo.[15] Bohm's other reference came from Oppenheimer, a fact that was later used in evidence against Oppenheimer during his security hearing.

Bohm's application was approved, and now all that remained was to obtain his passport. When his passport application was not processed promptly, however, he worried that the government might be restricting him. Einstein reassured him that he had discussed the matter with Oppenheimer, who felt that the United States had no intention of preventing Bohm's travel. "It might be that Oppenheimer takes the case too lightly but I too am inclined to think that nothing serious is behind the matter."[16]

Writing to thank Einstein for his help, Bohm also had good news about his theory: "It may interest you to know that Pauli has admitted the logical consistency of my interpretation of the quantum theory in a letter, but he still rejects the philosophy. He states that he does not believe in a theory that permits us even to conceive of a distinction between the observer's brain and the rest of the world."[17]

Finally, his passport approved, Bohm began to plan his trip to Brazil. When he met Oppenheimer on the street, his old mentor immediately remarked to him, "Haven't you gone yet?" At the time Bohm believed that Oppenheimer was giving him a friendly warning, suggesting that Bohm should leave the country as soon as possible. Several of Bohm's friends suspected an ulterior motive. By now Oppenheimer believed he too was to be the subject of a security investigation, in which case it would be better to have Bohm out of the country, where he could no longer be called on to testify against his former supervisor.

On the eve of his departure, Bohm returned to the subject of marriage with Hanna Loewy. An exchange of letters indicates the tension between them. Excited about the possibility of a career in Hollywood, Hanna gave up the idea of marrying Bohm "in any foreseeable future."[18] It was important for her to have the freedom to "find myself and to know where I stand in my work." After this rejection Bohm discussed his other women friends with Miriam Yevick and wondered if he should marry one of them so that he could have companionship in the new country.

Other clues to Bohm's thinking during this period come from a science fiction story he wrote. He left it with Hanna, along with other papers, when he flew to Brazil. While its metaphor refers to the anti-Communist feeling of the day, it also underlines Bohm's sense of alienation and his desire for the transcendent.

In the story, spaceships land on an Earth where life has become meaningless and cynical. After weeks of inaction, beings encased "in protective layers" emerge to probe the Earth people's thoughts. Then they begin to communicate, not telepathically but "through their deep understanding of the way the brain operates."

In the United States, politicians suspect that the landing is a Russian plot and drop hydrogen bombs on the spaceships—to no effect. Following this abortive attack, the Earth soldiers guarding the spaceships desert their posts and enter into more direct communication with the alien beings. The aliens tell them that Earth is not yet ready to become a member of an interplanetary culture. The human race has great potential, they say, but is at present like a baby who will one day grow into a great genius. The aliens' immediate concern is to prevent a war that could destroy the Earth.

While this message gives a sense of hope to people on Earth, a group of U.S. politicians and business people form a coalition to defeat the alien beings. They pretend to accede to popular demand and cooperate with the beings. But their aim is to learn some of their advanced scientific knowledge and use it to conquer the galaxy and spread the American way of life throughout the universe.

Despite their advanced knowledge of brain functioning, the aliens are of a somewhat overtrusting nature and readily agree to scientific collaboration with the Earth physicists. The Earth physicists learn from them that space and time are only approximations relative to "the absolute underlying the relative," allowing for speeds greater than light. But the more they probe into the aliens' science, the less they are able to understand.

Reactionary scientists working with the American coalition now believe that they are ready to launch an attack on the alien spaceships with new superweapons. The attack fails, since the beings are able "to control the very processes underlying these bombs." A wave of fear spreads across Earth as people anticipate a terrible retribution. The aliens, however, have always lived with the knowledge that they may be killed. They are sensitive and without any self-centeredness, "for a hostile thought would lead them to insanity." In fact, only by putting their whole being into their work were they even able to develop their science in the first place. Likewise, their knowledge can only be appreciated by those who have the proper emotional attitude toward life; for "the whole meaning of space, time and matter is found at the deepest levels to be inextricably bound up with the meaning of life and with the potentialities of intelligent life to work together for a common purpose."

The story ends with the aliens showing the human race how to share the power of thought, and a true socialist Earth comes into being. The aliens depart, planning to revisit Earth in a thousand years.

The political implications of Bohm's story are rather obvious; more interesting is the personification of many of his own yearnings in the form of "beings in protective clothing" whose science had developed in a wholehearted way and could not be understood by the average Earth scientist. Such a science transcends the exclusively material world, for it concerns the whole of life and consciousness.

In the 1970s, after having met the Indian teacher Jiddu Krishnamurti, Bohm revised the story (when relating it to the author) in a psychologically significant way. After the attack on their spaceships, the aliens are now plunged into a state of collective insanity. Despite their advanced science and absence of self, deeply buried forces that they had previously failed to acknowledge remain within their minds. Only by going into this dark area and transcending it are they finally able to emerge from the ships and transform humanity.

Bohm enjoyed telling this story to his friends and hoped that one day it would be developed into a film or novel. It is significant for the hints it contains of his own self-image—of someone without ego who is able to enter totally into the mysteries of nature to the point where consciousness itself is transformed and mind and matter become one. Later versions of the story acknowledge the existence of the shadow, that dark and destructive aspect deeply buried within each one of us. In this later version, Earth's final salvation lies in sacrifice, an encounter with the dark night of the soul and the emergence into light. This addition clearly referred to something in Bohm's own nature, something he sensed yet feared. Bohm had a desire to enter the unknown, yet he was repelled by its potential for chaos and irrationality. It was only in the last years of his life, in his final depression, that he appears to have fully faced this void. Tragically, by this time his health and energies were so frail that there could be no subsequent resurrection.

CHAPTER 8

Brazil: Into Exile

EXILE WAS A chilling experience. At Princeton Bohm had been surrounded by friends and supporters and supremely confident of his abilities. Photographs show a handsome, smiling young man, one certainly attractive to women. His work on plasmas was of international importance, his textbook on quantum theory was selling well, and to top it all, with his theory of hidden variables, he believed that he had made a major breakthrough in physics. Not only would this work revolutionize the way physicists thought about the world by stressing the rationality of nature, it would help to bring about a better society. Now he was being wrenched away from friends and colleagues to face years of isolation.

On the October day in 1951 when Bohm left the United States, thunderstorms and hurricane-force winds swept across New Jersey and New York. The flooded streets made it difficult for him to reach the airport, where Hanna and his father were waiting to see him off.[1] Nor was the weather Bohm's only concern. On board the aircraft while taxiing to takeoff, he heard an announcement that the plane would return to the terminal because of an irregularity in the passport of one of the passengers. Already highly nervous, Bohm wondered if he was going to be arrested. The problem, however, concerned another passenger, who was removed from the flight.

In the 1950s long air trips were undertaken in stages. Bohm's flight first landed in San Juan, Puerto Rico, where he found time to write a short note to Hanna Loewy mentioning that "the trip has been a long series of delays" and the frightening incident just before takeoff. By the time he reached Rio de Janeiro, he had recovered sufficiently to flirt with the "cute little girl" of about eighteen who escorted him to the waiting room. She asked if he knew any Portuguese, and when Bohm replied "not a word," she said, "But surely you must know the word for love—*amor*."[2]

On the evening of the second day, Bohm arrived in São Paulo where, to his dismay, no one met him at the airport. He had sent a telegram to the university telling them of his flight, but he had forgotten to address it to his new department. While the university boasted only a few thousand students, its various buildings were so widely distributed across the city that the cable had not been delivered.

The best Bohm could do was find a hotel and gesticulate until he was given a room for the night. Alone in a strange city, unable to speak the language, he fell into a state of panic. He could not even use the telephone since it was necessary to give the numbers in Portuguese. Eventually he calmed down a little and spent the night learning a few words of Portuguese from a book. On the following day, but only after considerable difficulty, he reached J. Tiomno, his contact at the university, who drove him to a pensione in 160 Avenida Angelica, where he was to live for the next few weeks.[3]

Bohm spent the following weeks in a heightened, labile state. At one moment he would plunge enthusiastically into his work and feel positive about his new life; at the next he would be dashed into depression and despair. At first the city appealed to him; he admired its buildings, and the pensione was comfortable. To Einstein he wrote that "it looks as if there will be an opportunity to do much good work." While the university was disorganized, "there are several good students here, with whom it will be good to work. . . . I have been continuing my work on the causal interpretation of the quantum theory, considering especially the Dirac relativistic wave equation. Here I have found that my interpretation leads to equations that seem to be closely related to the generalized theory of gravitation."[4]

Einstein replied: "Your letter has interested me very much. . . . I am very astonished about your announcement to establish some connection between the formalism of quantum theory and relativistic field theory. I must confess that I am not able to guess how such unification could be achieved."[5]

Nearing the end of his life (he died in 1955), Einstein was keenly aware of his failure to achieve the unified field theory he had sought for so long. In the general theory of relativity, he had taken an important first step toward unifying physics, by eliminating the force of gravity and replacing it with the curved geometry of space-time. He had hoped to do the same with the electromagnetic force, by extending the structure of space-time in some way. His dream was that quantum theory itself would emerge out of a final act of unification and that the universe would be described by a single theory (more specifically, by a single nonlinear differential equation). Bohm, however, was moving toward a very different insight, a universe of infinite levels, each qualitatively different from the levels above and below. Such a universe could never be reduced to a single all-embracing theory or equation.

Scientifically Bohm's first weeks in Brazil were exciting, but his initial impressions of São Paulo rapidly soured. He found the city full of confusion. Its ceaseless noise prevented him from sleeping. The traffic was dangerous. The climate was alternately too hot, too wet, or too cold. University politics was irritating, and the extreme contrast between poverty and wealth distressed him. It was not long before he discovered the sixth law of thermodynamics, valid only in Brazil: "Everything that is supposed to move is stationary, and everything that is supposed to be stationary, moves."[6]

Wherever he walked, he smelled the odor of decaying food, and it was difficult to find a restaurant where he felt sufficiently comfortable to eat. Inevitably he suffered from stomach problems and internal infections, which were to plague him throughout his time in Brazil. His diarrhea and fever became so severe that he was admitted to a hospital, where, thanks to a course of penicillin, his symptoms abated. Yet within days they returned—with such force that he became debilitated and "somewhat shaken in my convictions about the value of my work."[7]

There was one positive note. The southern hemisphere's long summer had begun, which meant that Bohm would not have to give lectures for several months. When his health permitted, he taught himself Portuguese, and he met a group of left-wing academics who made him feel at home. As to science, he pressed ahead with his research while waiting for his hidden variable papers to appear in print.

One problem he intended to solve with his new theory was that of the existence of two forms of statistics in quantum theory, called Fermi-Dirac and Bose-Einstein statistics. At the quantum level nature is divided into two distinct classes of particles. One class includes electrons, protons, and neutrons—the particles out of which matter is constructed. The second consists of particles responsible for the forces and fields of nature: photons and certain mesons. It turns out that particles in the former class have fractional spin—that is, the electron, proton, and neutron have a spin of ½—and obey what is known as Fermi-Dirac statistics, which means that only one particle can occupy a given quantum state. Particles in the second class have whole-number spins—the photon has a spin of 1—and are governed by Bose-Einstein statistics, which allow many particles into the same energy level or quantum state.

The connection between spin and statistics is one of the great mysteries of quantum theory, yet it has far-reaching practical implications. Without the restrictions imposed by Fermi-Dirac statistics, all the elementary particles in an atom could pack together into the same state. The entire earth, with all its trees, mountains, and people, would compact into a tiny, highly condensed, structureless blob. Thanks to Fermi-Dirac statistics, this compaction does not take place; rather, distinct chemical elements exist, and a wide variety of different compounds are possible. Conversely, Bose-Einstein statistics allow an enormous number of photons to pack into the same state, making lasers possible.

Why is nature constructed this way? Why do the elementary particles divide themselves so conveniently? Conventional quantum theory offered no explanation, and Bohm, helped by Tiomno, hoped that the solution would lie in his hidden variable theory. His 1952 papers described the motion of the electron (a Fermi-Dirac particle). He also attempted a theory of the electromagnetic field (its quantized particles, photons, are Bose-Einstein particles), but still the question remained of treating spin in a truly fundamental way.

In technical terms Bohm was able to deal with spin in the manner suggested by Pauli but not in Dirac's more fundamental approach. In between bouts of illness, Bohm was happily working on this problem with Tiomno until the young physicist moved to Rio. Now Bohm was stuck with the brutal fact of his isolation. The physics department showed little interest in his work, and his mail and scientific journals were taking far too long to arrive from the United States. Doing physics was food and drink to Bohm, but now that all external sources of stimulation had been removed, his ideas did not seem to flow as easily. His mood oscillated between two states: "In one I feel depressed and hopeless. In the other I just feel hopeless without being depressed. . . . Life is so tedious here that each morning, I wonder how I am going to get through the day. Under these conditions, my ideas tend to dry up, and my work just simply does not get anywhere."[8]

Things picked up for a time when he met another American, Phil Smith, who was "so full of life and at least he has good intentions."[9] The two men decided to share an apartment. While looking for a place to live, they ran into a difficulty; being euphemistically considered "bachelors," they were directed to the red-light district. But eventually they found a "beautiful apartment overlooking the city." Still, even with a place of their own, food remained a problem for Bohm, who found he could stave off the worst attacks of diarrhea only by eating in the most expensive restaurants. He did not turn to the obvious solution of cooking his own food since he "didn't know if I would be able to"![10]

Upon his arrival in Brazil, Bohm had been full of hope for a new life untainted by what he felt to be American-style cynicism and dishonesty. Now he realized that Brazil's economic dependence on the United States made its American ties particularly tight. A few weeks after his arrival, the American presence was brought home to him sharply. In late November or early December 1951, he received a visit from an official who requested that Bohm accompany him to the consulate for "registration and inspection of passports."[11] After handing over his passport, Bohm was told that it would be returned only if he flew back to the United States.*

Bohm immediately wrote to his friends in great distress. As he left

* Later, the legality of what had transpired became a matter of contention. Had the consular officials informed Bohm of his right to retain his passport and that he would continue to receive their protection as a U.S. citizen living in Brazil? According to Bohm, none of this had happened. His passport had simply been confiscated.

the consulate, he told them, "the people moved around with the comparative placid and unworried expressiveness of Brazilians, but it seemed as if the sun were being shaded by a faint cold hand that foretold the coming of the same kind of fear and tension that one sees everywhere in the US. It was really a shock to discover that I am not really out of the US."[12] The American authorities wanted to ensure that he would remain in Brazil,* but Bohm wondered if "something more serious is cooking."

On the following day he wrote to Miriam Yevick that "thinking things over last night, things do not seem quite as alarming." But he did believe he was being watched. Phil Smith told him that a car had circled the house for over an hour the evening before. "They probably wanted to see my reaction at having my passport taken away. I may assume that they will continue to watch me, at least to some extent."[13] It would be better, he decided, to avoid meeting colleagues who were sympathetic to the left.†

Despite his scientific work with Tiomno, Bohm's major concern during his first weeks in Brazil was to learn of the scientific community's verdict on his hidden variable paper, "An Interpretation in Terms of Hidden Variables." In January 1952, three months after his arrival in Brazil, he wrote, "It is hard to predict the reception of my article, but I am happy that in the long run it will have a big effect."[14] One concern did in the end prove lamentably correct: "What I am afraid of is that the big-shots will treat my article with a conspiracy of silence; perhaps implying privately to the smaller shots that while there is nothing demonstrably illogical about the article, it really is just a philosophical point, of no practical interest."[15]

One of the earliest reactions came from Bohm's good friend Richard Feynman, who had such a passion for bongo drumming that he attended a conference of the Brazilian Scientific Society in Belo

* There had been a fear that scientists possessing "atomic secrets" would find their way to countries behind the iron curtain. How the removal of a passport would have deterred a determined defector is not clear.

† Bohm's surveillance in Brazil, and later in Israel, would have been done under the auspices of the CIA. While researching this book, I applied for Bohm's CIA file under the Freedom of Information Act. The reply, which refused to admit that such files existed, indicated that even if these files did exist, they would not be released on grounds that included national security!

Horizonte as an excuse to spend time in South America.[16] The conference was a chance for Bohm to explain his new theory to Feynman. "When I met Feynman he thought that the idea was crazy but after enough talk I convinced him that it is logically consistent."[17] Feynman now agreed that "there might be something to my interpretation of quantum theory." He was "convinced that it is a logical possibility and it may lead to something new."[18] Bohm also hoped to shake his friend "out of his depressing trap of doing long and dreary calculations on a theory that is known to be of no use. Instead, maybe he can be gotten interested in speculating about new ideas, as he used to do, before Bethe and the rest of the calculators got hold of him."*

Bohm placed great weight on Feynman's reaction to his ideas. Several months after their meeting at Belo Horizonte, Bohm wrote, "Feynman was terrifically impressed with it, and now I think he is my friend for life."[19] He was right; even as late as the 1980s, he always visited Feynman during his trips to the United States. Bohm valued their discussions, saying (to the author), "Feynman is a very clever fellow," his highest compliment. Yet even though Feynman took Bohm's hidden variables seriously, he was not willing to work on the theory himself. The reason, Bohm explained, was that Feynman "could not see a problem in it."[20]

For Feynman, an area of physics was interesting only if it suggested a well-defined problem that could be tackled and solved. While Bohm and Feynman were always friends, a certain gulf separated their approach to science. Bohm had never been interested in the sorts of calculations that predict numerical results. For him, physics had to do with deep, relentless questioning and the exploration of fundamental ideas. Out of such an investigation would flow new physical insights. By contrast, Feynman saw physics as a challenge to his active, imaginative, and highly intuitive mind. His approach generally took the form of preoccupied wrestling with the puzzles and questions pre-

* Hans Bethe was noted for the discovery of the carbon cycle within the sun, an interlocking series of nuclear fusion reactions responsible for producing the heat and light of the sun and many stars. Bethe was later to develop aspects of the "Big Bang" theory of the origin of the universe, a theory that aroused Bohm's wrath and sarcasm because he believed it represented an unwarranted extrapolation of physics beyond its tested domain.

sented to him by nature. Probably Feynman would never have separated physics, as Bohm did, into deep and superficial realms, into fundamental questions and less important inquiries. Rather, he looked for those phenomena and questions that were intriguing, that exhibited a curious or unexpected side to nature, or that concealed, in some surface abnormality, an underlying unity.

It is difficult to say which approach serves science better. Important discoveries, like Einstein's formulation of relativity, come about through asking deep philosophical questions. Others, like Planck's discovery of the quantum, arise by seeking immediate solutions to puzzling problems (although Planck also believed that additional deep philosophical insights were needed before those immediate solutions could be integrated into the rest of physics). While Bohm attempted to wean Feynman from what he considered the "trap" of long calculations, one should not forget Eugene Gross believed that as a physicist, Bohm was sometimes lazy, preferring to ask deep questions rather than rolling up his sleeves and getting down to the nitty-gritty of calculations.[21] Bohm, of course, felt there was no point in doing such work. Clearly the best science is done when the two approaches complement each other.

If Feynman was supportive of Bohm's new interpretation, the same could not be said for other physicists. While still in Princeton, Bohm had received critical reactions to an earlier draft of the theory from Pauli. At the time he felt that he had successfully met these objections. Now he heard from Pauli again—he was "coming around a little (but just a *very* little)."[22] A few weeks later Pauli "practically concedes that the idea is logical."[23]

Then Pauli raised new philosophical objections. "I just received your long letter of 20 November," he wrote, "and I also have studied more thoroughly the details of your paper. I do not see any longer the possibility of any logical contradiction as long as your results agree completely with those of the usual wave mechanics and as long as no means is given to measure the values of your hidden parameters both in the measuring apparatus and in the observe [sic] system. As far as the whole matter stands now, your 'extra wave-mechanical predictions' are still a check, which cannot be cashed."[24]

Bohm disagreed—he believed that one day that hidden variables check could indeed be cashed. Pauli's metaphor concerned the utility of what Bohm had done. Since there was no way of detecting Bohm's hidden variables, he argued, then the theory offered nothing new. This was to be the common refrain in criticizing Bohm's theory. The predictions of conventional quantum theory are in remarkable agreement with experiment. If Bohm's theory had failed to retain that exact correspondence, then it would have been rejected immediately. On the other hand, if in every respect its predictions paralleled those of the conventional theory, then what was new about it? Why go to all the trouble of inventing additional assumptions when the existing theory worked perfectly well? In Pauli's opinion the underlying mechanism of Bohm's theory was truly hidden and, in this sense, an uncashable check.

Bohm was forced to agree that, as far as existing experiments went, the two theories agreed on all points and that nothing experimental could be said in favor of his own approach. But he looked forward to totally new experiments, ones carried out at very small distances and short time intervals, that might expose discrepancies with conventional quantum theory and decide the issue in favor of his own.

Pauli's letters to Bohm were polite and helpful, yet to his colleague Fierz, Pauli joked that Bohm kept writing him letters "such as might have some from a sectarian cleric trying to convert me particularly to de Broglie's old [1926–27] theory of the pilot wave."[25] In the last analysis Bohm's whole approach was "foolish simplicity," which "is of course beyond all help."[26]

Bohm had also sent de Broglie an early draft of his paper, and without waiting for the final published version, the French physicist had published objections in *Comptes Rendus*. Bohm replied in a letter to the *Physical Review*.[27] De Broglie, Bohm wrote to Miriam Yevick, had not "really read my article, but simply reiterated Pauli's criticisms, which led him to abandon the theory, but did not point out my conclusion that these objections are not valid. He's going to look a little silly. That's what he gets from rushing into print 5 months before my article came out."[28]

That was it as far as responses were concerned—the rest of the scientific community remained silent when Bohm's paper appeared in print. Bohm was convinced that he had achieved an important breakthrough, an entirely new approach to quantum theory, yet the scientific world was hardly beating a path to his door. As the days went on, he became increasingly frustrated and decided that his only tactic was to provoke the defenders of orthodoxy into attacking him. Such an attack would give him the right to reply and thereby initiate a debate that would be read by the general physics population. "I am hoping I can stir Pauli into writing a letter against it: this will stir up interest. . . . In this connection I am glad that de Broglie gives me an excuse to write a letter to the *Phys. Rev.*"[29]

The most important response of all would come from Niels Bohr, defender of the Copenhagen interpretation. Bohm waited with increasing anxiety as the weeks dragged on. Surely some word would come from Copenhagen. During this same period the philosopher Paul Feyerabend was visiting Bohr in Copenhagen. At the time he knew nothing of Bohm's hidden variable papers, but he had been impressed by the discussion of quantum measurement in *Quantum Theory*. "For the first time all this business about measurement made some kind of sense."[30]

The first Feyerabend heard of Bohm's new theory was during a seminar given by Niels Bohr. Following the lecture he asked Bohr to clarify certain points. The Danish physicist's reaction was, "Have you read Bohm"? As Feyerabend put it, "It seemed that, for him, the sky was falling in. . . .[31] Bohr was neither dismissive, nor shaken. He was *amazed*."[32]

In the midst of explaining to Feyerabend why Bohm's paper so disturbed him, Bohr was called away. The discussion continued without him for two more hours. Some of those present argued that the objections to Bohm's theory were not at all conclusive. As Feyerabend put it, the orthodox Copenhagen supporters tried to reply "in the Bohrian fashion." When this attempt was not successful, they said, "But von Neumann has proved . . . ," which ended the discussion. Feyerabend noted, however, that Bohr himself did not use von Neumann's supposed proof as a crutch in that fashion.[33]

What exactly was von Neumann supposed to have proved? In the early days of quantum theory, the mathematician John von Neumann presented a proof that quantum theory could never be reduced or transformed to any theory employing mechanical hidden variables.

It was the very proof that Bohm and Weinberg had often discussed at Berkeley. While most physicists had never bothered to read it, they paid lip service to von Neumann, assuming that his conclusion was true. Bohm, however, knew that the "proof" was based upon such restrictive assumptions that it did not rule out a hidden variables theory at all.

In Brazil Bohm knew nothing of Bohr's reaction, and his anger mounted at the total indifference of the physics community to an important new idea. It "cut at one's insides like a hot knife being twisted inside your heart."[34] He had a "passionate desire to fight this stupefying spirit of formalism, and pragmatism in physics,"[35] a spirit that focused on immediate, practical results while rejecting the underlying ideas as mere "window-dressing."

If Bohm never had news from Bohr, he did hear from Bohr's associate in Copenhagen, Leon Rosenfeld. The tone of that letter is somewhat like that of Saint Paul writing to an errant church: "I certainly shall not enter into any controversy with you or anybody else on the subject of complementarity, for the simple reason that there is not the slightest controversial point about it." Having asserted that he will not discuss the issue, he went on to address it "in the spirit of friendly conversation about some of the points you raise."[36]

Rejecting any possibility that the Copenhagen interpretation could be subject to reinterpretation, Rosenfeld admitted, might well cause Bohm to think that "my assertive attitude is extravagant"; for it is possible that he, Rosenfeld, and Bohr are also subject to error. But in their work together, Rosenfeld pointed out, he and Bohr had already made all the errors that could conceivably be made before arriving on solid ground: "It is just because we have undergone this process of purification through error that we feel so sure of our results." Therefore, when Rosenfeld made assertions of infallibility, he said, he was not being dogmatic, for "there is no truth in your suspicion that we may just be talking ourselves into complementarity by a kind of magical incantation. I am inclined to return that it is just among your Parisian admirers that I notice some disquieting signs of primitive mentality.[37]

"I find your position regarding 'simplicity' a bit confusing," Rosenfeld continued. "You start by telling me that the causal point of view is to be preferred because it is simpler, but further on you state that those who reject it are oversimplifying the issue." Simplicity could be of no use as a criterion for the suitability of a physical theory, he asserted—physics had to adapt its thinking to nature, whether this process was simple or not. "The main thing is not to accept any other guidance than that of Nature herself."[38]

Annoyed, Bohm felt Rosenfeld had completely missed the point. To Miriam Yevick, he compared Rosenfeld to Ernst Mach who, at the end of the nineteenth century, had attacked the early atomic theory on the grounds that, when it came to explaining the behavior of gases, the theory produced results identical to those of classical mechanics. There is no need, Mach had argued, to invoke the additional assumption of atoms. "Rosenfeld is a confirmed positivist," Bohm continued, "so deeply confirmed that like M. Jourdain, who spoke prose without knowing it, they [*sic*] speak positivism without knowing it and call it dialectics."*[39]

As the weeks went by, Bohm lashed out not only at the indifferent physics community but at friends he felt were turning against him. One of his outbursts was precipitated when Miriam Yevick mentioned in a letter that Eugene Gross had banged on the table, saying that Bohm had to get results. Bohm felt that "one of my best friends seemed to be turning away from me, and running with the tide."[40] He despised contemporary physics for its insistence on results. Plenty of trivial results had appeared over the last twenty years, but no real advances, he thought. People like Julian Schwinger and Richard Feynman may have produced "resultlets," but "these little mice are all that come out of twenty years of labor by the mountain of thousands of theoretical physics." Now Bohm was being asked "to give them results . . .[41] I alone am supposed in a year or two to produce a scientific revolution comparable to that of Newton, Einstein, Schrödinger and Dirac all rolled into one."[42]

When another erstwhile supporter defected, it was "comparable to a man who shoots you in the back, and then begs pardon saying that he was using a theorem from which he deduced that bullets come

* In Molière's play *Le Bourgeois Gentilhomme* M. Jourdain is ironically flattered when told that the language he is speaking is prose.

out from a gun at an angle of 90° relative to the barrel, and that he was therefore really shooting at someone else."[43]

When reactions did occur, they were hardly what Bohm had hoped for. While "von Neumann thinks the idea consistent, and even 'very elegant' (The unprincipled bum!),"[44] Niels Bohr had told the physicist A. Weightman that Bohm's theory is "very foolish."[45] Leon Rosenfeld wrote to a friend saying that the theory is "very ingenious, but basically wrong."[46] As for Erwin Schrödinger, he "did not deign to write me himself, but he deigned to let his secretary tell me that His Eminence feels that it is irrelevant that mechanical models can be found for the quantum theory, since these models cannot include the mathematical transformation theory, which everyone knows is the real heart of quantum theory. Of course, His Eminence did not find it necessary to read my papers, where it is explicitly pointed out that my model not only explains the results of this transformation theory, but also points out the limitations of this theory to the special case where the equations are linear. . . . In Portuguese, I would call Schrödinger *un burro*, and I leave it for you to guess the translation."[47]

Neither, finally, did Einstein favor the theory. While he still believed that orthodox quantum theory was incomplete, he felt that Bohm had "got his results too cheap." "This path seems to me too easy," he wrote to Max Born.[48] It was "a physical fairy-tale for children, which has rather misled Bohm and de Broglie."[49]

Einstein also stated his objections in a letter to Bohm's former student Daniel Lipkin.[50]

> I too have many reasons to believe that the present quantum theory, in spite of its many successes, is far from the truth. This theory reminds me a little of the system of delusion of an exceedingly intelligent paranoiac concocted of incoherent elements of thought. As you also seem to believe as I believe it impossible to get a real insight without satisfying from the start the principle of general relativity. I feel, however, by no means sure that my own approach is the right one.
>
> I do also not believe that the de Broglie–Bohm's approach is very hopeful. It leads, f.i., to the consequence that a particle belonging to a standing wave has no speed. This is contrary to the well-

founded conviction that a nearly free particle should approximately behave according to classical mechanics.

While he was waiting for his theory to appear in print, Bohm had worried that the "big-shots" would greet it with a conspiracy of silence. Unknown to him, this is exactly what was happening, at least in the experience of Max Dresden, a physicist at the University of Kansas. Several of Dresden's students read Bohm's *Physical Review* paper and asked their teacher what he thought about it. Dresden's first reaction was that von Neumann had proved that hidden variables do not exist. This did not deter his students, who formed a Bohm study group.

In the end Dresden was forced to read Bohm's paper. He had assumed that there was an error in its arguments, but errors proved difficult to detect. He also read von Neumann's "proof" and realized that it did not rule out the sort of theory Bohm had proposed. In January Dresden visited Oppenheimer and asked his opinion of Bohm's theory. "We consider it juvenile deviationism," Oppenheimer replied. No, no one had actually read the paper—"we don't waste our time." Dresden said that he was troubled by the issue and did not know what to make of it. Oppenheimer proposed that Dresden present Bohm's work in a seminar to the Princeton Institute, which Dresden did.

The reception he received came as a considerable shock to Dresden. Reactions to the theory were based less on scientific grounds than on accusations that Bohm was a fellow traveler, a Trotskyite, and a traitor. It was suggested that Dresden himself was stupid to take Bohm's ideas seriously.

Included among the accusations were scientific objections that Dresden found difficult to answer. But all in all the overall reaction was that the scientific community should "pay no attention to Bohm's work." As Dresden recalled, Abraham Pais also used the term "juvenile deviationism." Another physicist said Bohm was "a public nuisance." Oppenheimer went so far as to suggest that "if we cannot disprove Bohm, then we must agree to ignore him."[51]

A hint of the Princeton reaction must have reached Bohm in Brazil, for writing to Miriam Yevick, he lashed out in frustration: "As for Pais and the rest of the 'Princetitute' what those little farts think is of no consequence to me. In the past 6 years, almost no work at all has come out of that place. . . . I am convinced that I am on the right track."[52]

The Princeton Institute's reaction is particularly disturbing since it was motivated as much by political expedience as by scientific and philosophical objections. It did not surprise Ellen Schrecker, a historian from Yeshiva University whose *No Ivory Tower* was a special study of academics during the McCarthy period.[53] In other cases, too, she found, academics had formally distanced themselves from colleagues who were politically "tainted." The 1950s version of "political correctness" required not only stating one's support for the contemporary political status quo but actively disparaging fellow travelers and the politically suspect.[54] A physicist and colleague of Einstein, Peter Bergman, went so far as to use the term "ethnic cleansing" in reference to this period. In his opinion the Princeton Institute was "infected by a number of prima donnas," Oppenheimer among them. It was extremely difficult, he said, to separate scientific reactions to Bohm from the need to dissociate from a politically tainted former colleague.[55]

J. Robert Oppenheimer's role in effectively silencing Bohm's work is particularly complex. By the early 1950s, Oppenheimer had made powerful enemies, and he feared that a security investigation would be directed against him. During the war Edward Teller had argued forcefully that the United States should develop a hydrogen bomb (fusion bomb) instead of the atom bomb (fission bomb) that the Manhattan Project eventually built. At the time Oppenheimer had opposed Teller. The allegation could now be made that Oppenheimer had compromised national security by delaying the development of the more powerful weapon.

According to his former friend and associate, Melba Phillips, Oppenheimer was concerned not only with his personal survival but with the future of atomic energy. Never modest even at the best of times, Oppenheimer identified himself with the future of this new energy source and the transformation of the world by limitless energy. He was the only individual with sufficient authority and vision, he believed, to bring about international collaboration. At all costs he must survive, even if scapegoats had to be called upon to pay a price on his behalf.[56]

Oppenheimer therefore had particularly strong reasons to distance himself from a young man who had not only fled the United States under a political cloud but was now attacking the very foundations of quantum theory. At a time when the Copenhagen orthodoxy was analogous to that of Viennese psychoanalysis, in which defection was taken as the ultimate betrayal of revealed truth, Bohm's apostasy must have been shocking and distasteful to Oppenheimer.

To the present reader, it may be surprising that a scientist would have used Bohm's politics as a stick to beat his physics. Yet in the 1950s Bohm himself made no distinction between his work in physics and his political and philosophical beliefs. Human freedom and social transformation were, in his mind, tied to how people understood the material world. The metaphysical confusions introduced by quantum mechanics acted as a barrier to the development of a more rational society. If only people could understand that causality operated right down to the level of the atom, then they would begin to apply it in their own lives and behave in more rational ways.

In the early 1950s, in any event, word appears to have gone out, either directly or tacitly, that the leaders of the scientific community were ignoring Bohm's hidden variable theory. That this reaction was picked up by those lower down in the scientific pecking order is the only explanation for the conspiracy of silence that greeted Bohm's papers. Since 1952 many other interpretations to quantum theory have been proposed. Some are quite bizarre, yet most find their way into textbooks and scientific reviews. By contrast, until the 1990s, virtually no mention was made of Bohm's hidden variables. As Feyarabend put it, "The fact . . . that Bohm's model was pushed aside while all sorts of weird ideas flourished is very interesting, and I hope that one fine day a historian or sociologist of science takes a close look at the matter."[57]

Take one example of the weight of this silence: At Temple University at a conference on quantum theory in the 1980s, Bohm spoke about the latest developments in his theory. The meeting ended with a survey of the various interpretations of quantum theory, given by Hillary Putnam, a well-known philosopher of science. Even though Bohm was sitting in the front row, no reference was made to his approach.

Desperate to confront this wall of silence, Bohm longed to travel to Europe and the United States, where he could give seminars about his work and challenge his detractors. But without a passport it was impossible for him to leave Brazil. Had he really done the right thing in leaving the United States?

But as news of the hardening American political climate reached him, he realized that he had made the right decision and resolved to "return to the States only if they dragged me back." He despised the extreme pragmatism that he believed had led the American people to their present state: "I have hated this spirit, even as a child, especially since it was so strongly embodied in my father."[58] And: "As far as my hatred of America is concerned, you must remember that this is because I once loved it very much, as the land where there was freedom from the ties of the past that bound Europe, and where there was hope for the future."[59] This dream would never come to fruition, he concluded; "the correct emblem for the U.S. should be, not the eagle, but a pig with his nose in the pork-barrel." What supports a culture of "cynicism and cowardice," he wondered, one that leads younger physicists "to accept without protest whatever the "big-shots" tell them"?[60]

Two years later Bohm's hostility had not abated. "I seem to have only one strong emotion left—and that is hatred for the forces that have destroyed so many human beings, including myself. For relative to what I could have been, I regard myself as destroyed." He can never forgive the "American Way of Life," its selfish, cold calculation, and the conformity and insincerity that stultifies everything that is not concerned with profit. "I have hated the American Way of Life from the moment in which I was conscious of the need to take care of myself against others, from the moment in which, as a child, I had to hide books, so that people would not think I was 'abnormal' or a 'sissy.' . . . I cannot forgive them for creating conditions in which I could have so few real friends, and in which I lost the opportunity to work together in a corroborative way with other people towards a common goal that was worthwhile. I cannot forgive them for making it necessary for me to live a life in loneliness and futility. . . . I cannot forgive them for turning most people into dull, limited creatures that they are."[61]

His country gone to the dogs, his theory ignored, only his faith could keep Bohm going—a faith in the combination of science and Marxism. "People's ideas are in the long run fundamentally influenced by their concepts of the fundamental properties of matter," he wrote, "and the idea that the properties of matter are understandable in a natural way as well of much interest and intellectual beauty, will help move people's ideas away from the confusion and mysticism in which they are now mixed up."[62] People adopt a cynical attitude toward life when they assume that their morality and responsibilities have no objective basis and are determined by the prevailing society. The materialistic view shows that "morals grow out of the material structure of human beings and out of the material relations between human beings and society.[63] . . . Unless the material basis of life is understood and controlled, life cannot attain its full potential."[64] Moreover, "if people do not try to satisfy their curiosity and increase their control of nature, they stifle an important part of themselves and cannot become complete human beings."[65]

Bohm's refuge from disappointment, frustration, and internal darkness was the world of ideas. Hidden variables had not been accepted, but he had to press on. That first Brazilian summer, he began to make new speculations about elementary particles. At Berkeley he had pictured the electron as a wave that constantly collapses inward from all space and expands outward again. Now he began to develop a picture of the electron based on his theory of plasmas. At Princeton he had shown how, within the random motion of electrons, the collective behavior of the entire electron gas is enfolded—the plasma vibrations. In Brazil he discovered an entirely new type of collective behavior, one that was not spread over all space like the plasma but localized into a small region. In this description what looked like particles were in fact the collective albeit localized behavior of the astronomical numbers of electrons moving in a "highly interconnected" way.*

Again, the vortex that had fascinated the young David Bohm, formed when water runs down the drain hole of a bath, is illustrative.

* Bohm had discovered what are today called the "elementary excitations" of a metal. Much of solid-state physics is now expressed in terms of these "elementary excitations."

The vortex, a stable entity, is the manifestation of the movement of a large amount of water. What looked like localized entities in a metal, he now saw, could be the collective effect of a very large number of electrons. It was a short leap of the imagination to suggest that elementary particles themselves could be the result of astronomical numbers of even smaller particles behaving in a collective way. What if all of space is "made up of particles millions of times smaller than the electron or a proton"? he asked. This idea dates back to his high school theory of a four-dimensional ether.[66]

Experiments with elementary particle accelerators are interpreted in terms of electrons, protons, neutrons, mesons, and so on. One day, Bohm speculated, experiments carried out at much smaller distances will reveal further levels of matter and new and far smaller particles. These microlevels will, in turn, be merely stages in an endless series of ever smaller levels. In Bohm's vision, the individual particle is conditioned by the collective, while in turn, the higher levels help "determine the character of things that may exist at the lower levels."[67]

At Princeton Bohm had read an article by Lenin, arguing that the electron is inexhaustible in its possibilities. This notion may have flown in the face of modern physics, but it struck a chord with Bohm. Now he was able to describe the electron as a complex structure moving within an underlying ether. At distances "millions of times smaller" than the electron, Einstein's relativity would break down to reveal "absolute space"—an ether of tiny particles.*[68]

Over the following weeks Bohm developed the details of his new theory. He showed that an electron in a metal is a little like a speedboat in a lake. As it moves, the electron is surrounded by a spreading cloud of charge. The faster it moves, the more the cloud bunches up in front and acts to slow down the particle. The limiting speed of an electron, or any material particle, is determined by the speed at which

* Several decades later Bohm interpreted the Lorentz transformations between different moving frames—the backbone of special relativity—in terms of transformations between what he termed "material frames." These material frames correspond in certain ways to the complex collective structures in an underlying medium that Bohm began to explore in the early 1950s. Until the end of his life Bohm actively pursued this approach to resolving quantum theory and relativity.

this cloud, or wave, spreads. In fact, it turns out to be the speed of light. Try to accelerate the electron beyond the speed of light, and the bunched-up waves immediately slow it down. Just as Einstein had predicted from his relativity theory, the speed of light is the limiting velocity for matter.

Bohm preferred his own theory to Einstein's because it gave a material reason for the speed of light as a limit and, at the same time, allowed for exceptional faster-than-light motion (something not permitted by Einstein). Suppose the electron is given a sudden, violent impact. For a short time it will be ripped free of its cloud and free to travel at an unlimited speed. This is only a temporary condition, and it will soon be slowed down again.[69]

In his excitement Bohm wrote to Melba Phillips, "I have become convinced that the time has come to reconsider the concept of an 'ether' that fills all space."[70] He pointed out that "there has never been a proof of the non-existence of the 'ether.' " Physicists generally assume that the famous Michelson-Morley experiments, carried out at the turn of the century, demolished the ether and paved the way for Einstein's special theory of relativity. But all these experiments implied, Bohm now pointed out, is that the present laws of physics can be written without reference to the ether. They did not rule out that the ether could manifest itself in some new domain.

Using this ether, Bohm would reintroduce the idea of absolute space, which Einstein had rejected. Yet even absolute space was only "relatively absolute," for "its particles are in turn made of something still smaller, etc., ad infinitum." His conclusion was that "all matter contains an infinity of qualitatively different levels, all interconnected."

Bohm's new ideas were a return neither to reductionism nor to mechanistic thinking, he was at pains to point out. The reductionist approach seeks to explain phenomena in terms of an elementary and underlying level. To Bohm, the electron is determined by an underlying substratum, yet the tiny particles that make up this substratum are themselves partly determined by their participation in overall collective movements. An analogy may help at this point. The human body is composed of organs that are, in turn, built out of cells. The cells of the brain, heart, and liver all look quite different and reflect the functions that these respective organs perform within the body. While organs are built out of cells, the form and function of these cells is conditioned by their presence within a particular organ. The human body is an integrated, living system not reducible to interlocking mechanical parts. In a similar way Bohm came to view the universe as a holistic, organic structure.

Even if the universe is a living thing, however, matter still reigns supreme. Particular material forms, such as electrons or protons, come into and go out of existence (as a result of transformations at a lower level), but "only matter as a whole, in its infinity of properties and possibilities, is eternal." Physics, by contrast, is provisional: "I believe that no law is absolute or final, but that each law provides a successively better approximation to an absolute truth, that we can never possess in a finite time, because it is infinite in all of its aspects, both qualitative and quantitative."[71]

Through "work and study," nonetheless, Bohm felt it possible to uncover level after level and in this way move closer to "absolute understanding." "Humanity as a whole (combined with other forms of intelligent life) has the possibility of exploring knowledge without limit." "Understanding the causal laws" will allow human beings "to go beyond all conceivable limitations. In this fashion a true understanding of causality and necessity would bring about human freedom. Failure to do so would result in our enslavement by external nature and our own natures."[72]

Bohm's meditations on causality, determinism, and chance evolved into one of his most important books of the period, *Causality and Chance in Modern Physics*. Already he had begun to explore the book's key arguments in letters to his friends. By causality, he explained, he meant that a knowledge of causes enables science to predict effects, and that by changing the causes, it is possible to change the effects in a predictable way.

At first sight, this idea involves a paradox, since absolute determinism would clearly make it impossible to change anything at all. In a causal world that contains only a finite number of levels, all is preordained. Complete causality implies complete determinism and no possibility of freedom.

With Bohm's qualitative infinity of levels, however, the two concepts of determinism and causality moved apart. While there is complete causality at every level—in the sense that every effect is the direct result of a cause—with an infinite number of qualitatively different levels the world is not completely determined, and the emergence of the qualitatively new is always possible. Effects from a limitless number of lower levels surge up to a higher level to produce qualitative changes that cannot be described in terms of what already exists at that level.

Bohm's metaphysical vision was of an endless dialectical process, one that proceeds until it encounters a contradiction that can be resolved only by going to the next level. This process, Bohm pointed out, avoids the nightmare of complete determinism. While a given level is causal, the overall effect of the totality of levels, each one qualitatively different from the others, can never be fully taken into account. "Thus, *as a matter of principle,* we say that complete determinism could not even be conceived of. Yet, each level can be determined." Bohm's infinity of levels, together with his new picture of the electron, is therefore neither deterministic nor reductionistic.

Dialectical materialism, for Bohm, was a necessity that arose out of the very structure of matter. Now, with the aid of his infinity of levels, he could give it a particular and clear expression. For a time he even planned to write a book on dialectical materialism, but he became discouraged when he thought of the "positivistic and idealistic baggage" that even his friends on the left carried. He remembered a "violent" argument with George and Miriam Yevick who had suggested that the dialectic could be based on several opposing forces and shades of distinction. This was the sort of corruption people were picking up from a "bourgeois society" that focused on the superficial and on shades of difference. The bourgeois always accept situations as fixed, with only limited possibilities for change.

Bohm selected Aldous Huxley's *Brave New World* as a particularly odious example of this tendency. Huxley's fashionable ideas "stink" because of his bourgeois tendency to crystallize the contradictions that exist within present society and to regard them as inherent in human nature. By complaining about such situations, people feel excused from having to do anything about them. For Bohm, it was always necessary to dig deeper in order to discover the underlying forces that shape human nature itself.[73]

For a time Bohm thought about writing an article on hidden variables that would make an explicit connection to his political ideals. When Miriam Yevick met with I. F. Stone, who was planning a socialist journal, he suggested it would be a venue for the essay. At this prospect Bohm got cold feet, worrying that the essay could be seen by "someone of influence" who would demand that he return to the United States: "Without a passport, I am a sitting duck for such an attack."[74]

In a period when his science seemed to have been rejected, Bohm's political beliefs sustained him. Indeed, as one who lived for ideas, he was always in danger of being possessed by them. Sometimes the results could be chilling, as when he hit out at "middle-class leftist" friends of the Yevicks who had not been pleased with what they had seen in Yugoslavia. Such people sell out when the going gets tough, he said, and "dream of socialism as a fine way to do as they please in return for 4 hours a day of work."[75] Multiply their attitude by hundreds of millions of people, and socialism becomes an impossibility. Such an attitude could be combated only by the most energetic of measures, which meant measures that may seem "more energetic than is absolutely necessary." The alternative is the collapse of the entire society. For this reason Bohm could understand "if things are done with excessive crudeness" when there was insufficient time for a smooth transition.[76]

Nevertheless, sometimes he did worry about reports coming from the Soviet bloc: "I am getting scared a bit. Everything seems so mixed up,"[77] for "tremendous mistakes were made in the USSR and will doubtless be made in the future."[78] However, "these things cannot alter the fact that in the long run, things will come out O.K." Only Marxism had the power to transform society and create greater human freedom.

Eager for more information on the Soviet Union and the rest of the Communist world, he asked Miriam Yevick to order some books for him from Lawrence and Wishart in London. She should also order some for herself, he suggested, in order to "counterbalance the heavy diet of Koestlerian literature you have been absorbing lately."[79] His order (quoted here in full) gives an indication of his reading at that period:

1. *Art and Social Life* G. V. Plekhanov 25s
2. *China's Feet Unbound* W. G. Burchett 5s
3. *The Soviet Union Today (A Scientist's Impressions)* S. M. Nanton F.R.S. 9s 6d
4. *How Music Expresses Ideas* Sidney Finkelstein 9s 6d
5. *The English Rising of 1381* R. H. Hilton and T. H. Aston 10s 6d
6. *Francis Bacon, Philosopher of Industrial Science* Benjamin Farrington 12s 6d
7. *Life and Teaching of Friedrich Engels* Selda K. Coates 1s
8. *On Contradiction* Mao Tse-tung 4d
9. *The True Story of Ah Q'* Hsun Lu 1s 6d
10. *I Fought in Korea* Julian Tunstall 8s 6d.

Appendix

Within a Marxist environment any deviation from materialism, any taint of mysticism, was to be roundly condemned. Bohm's own polemic was directed against what he felt to be the combination of mysticism and unwarranted scientific extrapolation inherent in the Big Bang cosmology advocated by George Gamow and his co-workers. The ultimate origin of the universe, if indeed it has an origin in space and in time, is one of the major questions asked by all cultures. The question was asked anew after 1929, when astronomer Edwin Hubble discovered that the shift in the color of the light coming from distant galaxies indicated that they were receding from each other at high speeds. The universe, Hubble demonstrated, is rapidly expanding. Extrapolating back in time for tens of billions of years, the cosmos must have originally been compressed into a very tiny region of space. Such a possibility was allowed in Einstein's general theory of relativity.

Scientists began to speculate that far back in time the universe consisted of matter and energy of enormously high densities and temperatures. But these early theories kept running into technical difficulties. Finally Gamow and his co-workers helped to develop the theory of the Big Bang.* (Gamow was also the author of a minor masterpiece of science popularization, *Mr. Tomkins in Wonderland*.)

When Brazil's popular press got hold of the big bang theory, it was presented as the confirmation of the biblical account of creation. Bohm came across a headline that read, "The Existence of God Proved Scientifically." "And how was this miracle accomplished and by whom?" he wrote to Miriam Yevick. "Apparently by none other than G. Gamow, who works for George Washington University, which is, I believe a Catholic Institution."[80]

Gamow's theory was "almost pure speculation . . . a pure web of fantasy," Bohm told Miriam. His sarcasm was also directed at the pope, who had used the theory to deduce that the beginning of time is beyond human understanding, "thus creating a new mystery paralleling the Trinity in its fathomless depths." However, "in the midst of all this beatific satisfaction with a new reconciliation of science and religion," there were, Bohm felt, weak points in Gamow's argument. Gamow had simply extrapolated laws of nature, deduced at the low density in which humans exist and design their experiments, into a hypothetical period of very high density when the universe was compressed into a very tiny region. At such densities, Bohm argued in his letters, "as yet unexpected levels may come into play." Nevertheless, "as expected from reactionaries," Gamow extrapolates his conceptual structures into new domains and thereby creates new mysteries.[81] Gamow had weakened the whole approach of materialism, since for him, "matter begins to take on a shadowy character of ideas. In other words, he [the pope] dematerializes the concept of matter."

Bohm admitted that the pope was a skillful tactician, "fighting a desperate losing battle to save the Catholic Church." Science, with its great prestige, is ideologically the principal enemy of religion, "which

* Over subsequent decades the Big Bang theory underwent a number of modifications, such as the "inflationary model," but in its general form it remains the explanation favored by most scientists. More recently, however, the notion of a single instant of creation has become more controversial.

latter is of course ideologically one of the main props of the existing order." The pope therefore "welcomes scientists who effectively turn traitor to science, and discard scientific facts to reach conclusions that are convenient to the Catholic Church."

The strength of Bohm's attack, particularly that matter would take on "the shadowy character of ideas," is particularly ironic when seen in the light of his later attraction for Hegel and the teachings of Jiddu Krishnamurti.

CHAPTER 9
Causality and Chance

WITH THE BRAZILIAN SUMMER of 1952 over, Bohm began to lecture in Portuguese. Over the previous months he had slipped into a leisurely routine of half an hour's scientific reading after breakfast before getting down to physics. After lunch he took a short nap that within a year graduated into a full two-hour siesta. As the afternoon cooled, he would walk, returning home around six to rest before going out to eat. He spent the remainder of his evening writing letters.

Half an hour's reading a day is a particularly small amount. "Keeping up with the literature" is a major task that requires setting aside many hours each week. But unlike other physicists, Bohm never gave much attention to the scientific journals. Most of what was published seemed of little value to him. When a significant advance did occur, he preferred to talk it over with a colleague and extract the essence of the new idea—generally transforming it in the process.

Bohm's indifference to scientific papers meant that he gave few references to other people's work in his own published papers. It was not that he was stealing other people's ideas or claiming them as his own; rather, he simply could not be bothered to spend time tracking an idea down to its source. Those few ideas that did fascinate him tended to become so integrated into his own thinking that he would often forget their origin.

During his first academic term Bohm confronted a university administration grown chaotic in the wake of World War II. The physics department was in such bad shape that its students were forced to take many of their courses in the mathematics department. Nonetheless, the professors continued to work independently of each other, with no apparent regard for their students' needs. Bohm planned to change all this by teaching properly designed courses. He was also set to do battle with the unwritten agreement that students would first attend a class around a month after it had officially commenced! Despite the self-deprecation of his letters, Bohm appears to have been successful in inspiring his colleagues, for later, after he moved on from Brazil, he left behind an active and stimulating department.

Bohm made contact with other left-wing intellectuals in the university, but he was troubled by Brazil's pervasive anti-Communist attitude. Shortly before his arrival a well-connected student had gone so far as to ask the army to intervene in the physics department because of the number of Communists present. While this intervention never materialized, Bohm grew uneasy when the same young man became involved in a proposed Institute of Theoretical Physics in São Paulo, for which $200,000 had been obtained from the state government and $800,000 from private sources. The fly in the ointment was that Werner Heisenberg and [name withheld] were to be co-directors, working in three-month shifts. Bohm's fear was that they would let "the rest of the Nazi Vermin in."[1]

Bohm opposed the venture, and early in May 1952, he discovered that "all hell is ready to break loose"[2] since "this Nazi group has joined up with the local dep't stinkers, and found out about my past from Europe." When Bohm tried to get rid of the well-connected young man, the "stinkers" responded by suggesting that Bohm should take German assistants in place of the young American, Ralph Schiller, who was about to work with him. If Bohm agreed, he was ominously warned, "we could avoid the department's reputation as 'communist' and thus avoid army intervention."

Bohm was prepared to fight back by letting "the world know what a dirty stinker [name withheld] is."[3] Writing to Melba Phillips, he said, "I would like to find some way to let the world know what a skunk [name withheld] is, or at least, let all the physicists know . . . it might make a good story 'Nazi's taking over Brazilian Physics' (We are only 90% sure of [name withheld]'s role in all this)." Bohm contacted Einstein and asked Phillips to inform his friend, the physicist Philip Morrison, and he wrote to Miriam again: "try to see what you can do about lining up publicity against [name withheld], *but don't do a thing till I say 'go.'* " He also asked for a copy of a scientific proof: "it is important in the struggle against [name withheld] & the rest of his rats, as they are trying to attack my scientific prestige in comparison to his. The dep't stinkers go around calling him a 'genius' every time he opens his mouth."[4]

Matters became so serious that the head of the faculty, Abrahão de Moraes, wrote to Einstein: "As a result of the difficulties that Professor Bohm experienced in the United States there exist possibilities that he may have analogous difficulties here." Would Einstein write a letter suitable for the press and others addressed to "Governor of the State of S Paulo, Dr. Lucas Nogueira Garcez, and to the President of the Republic, Dr. Getulio Vargas"?[5]

On May 24 Einstein replied, "Dr. Bohm, whom I have known personally for some years, is, in my opinion, a very gifted and original theoretical physicist. Professionally, he has added materially to our knowledge of quantum mechanics and has more recently become very interested in the fundamental philosophical implications of that theory. He is also an exceptionally able teacher who is an inspiration to his students." As for Bohm's political difficulties, Einstein said they were the result of the "tense situation after the war, [and] have no bearing whatsoever on Dr. Bohm's moral character."[6]

In the letter intended for publication, Einstein wrote that Bohm, "having already achieved a very stimulating book on Quantum Mechanics and its application to the theory of atoms, . . . has become deeply interested in the following questions. Is it really necessary to assume that the processes in the molecular domain are governed by chance? Is it not possible to explain the present theory in such a way as to indicate that everything should proceed by necessity, so that chance is, in principle, eliminated. . . . To me, Dr. Bohm has a lovable personality, and with this opinion his former students and colleagues would agree wholeheartedly. . . . I have had in the past the greatest confidence in Dr. Bohm as a scientist and as a man, and I continue to do so."[7]

The situation in the physics department, combined with construction noise that was robbing him of sleep and news of increasing U.S. involvement in the Korean war to the point where use of atomic weapons was being considered, coalesced in Bohm's mind into a crisis. He wondered if "the game is up for everything living in this generation, and this largely because of atomic power, which seemed such a bright dream when I was a child."[8] His only hope was that his new ideas in physics would somehow compensate for what was being done by the atomic bomb.

In the end the confrontation with the physics department blew over. A new institute was opened "at an official ceremony full of generals, colonels,"[9] but [name withheld] and Heisenberg were not involved. Bohm could now turn his energies back to research with his new American assistant, Ralph Schiller. His old friend George Yevick also arrived on a scientific visit, and his cheerful, optimistic nature gave Bohm such new confidence that he even planned to write a book on dialectical materialism.[10]

All in all, things were beginning to look up for Bohm. Richard Feynman arrived for the bongo drumming, and he, Yevick, and Bohm spent several days at Copacabana Beach. In the light of Feynman's considerable reputation, Yevick was surprised that Bohm took the role of intellectual leader, doing most of the talking while Feynman walked beside him asking questions.[11]

Before arriving in Brazil, Yevick had been to Paris to lecture on Bohm's hidden variable theory. Louis de Broglie was not present, but later, when his assistant, Jean-Paul Vigier, explained to him the new development of Bohm's ideas, the older physicist (at least according to Vigier) became so excited that "he jumped on a chair in excitement." This report convinced de Broglie to revive his own discarded theory, and he sent Vigier to Brazil to find out firsthand what was going on.[12]

Like George Yevick, Vigier had intense enthusiasm and a childlike innocence. Both men acted as an anodyne to Bohm's retiring nature. Unlike Yevick, however, Vigier was a Marxist and took his politics seriously. In the Second World War he had been on the general staff of the French Resistance. Once after being captured and beaten, he was sent to the personal attention of Klaus Barbie, the "Butcher of Lyons," and only escaped thanks to a Royal Air Force bombing raid on the train. Later he became a personal friend of Ho Chi Minh.

In Brazil Vigier and Bohm sought to answer one of the major objections that had been made to the hidden variable theory, namely that it did nothing new—the jibe of Pauli's that Bohm's check was uncashable. Their idea was to extend the theory so that its predictions could be tested against those of conventional quantum theory. The case they chose was that of probability. Conventional quantum theory claims that probabilities are absolute and irreducible to any underlying explanation. The result of one measurement is quite random and uncorrelated when compared with one made only a moment later. No amount of experimental manipulation or control can ever alter this state of affairs.

By contrast, Bohm's theory did not necessarily accept absolute probabilities, for what looked like chance results were in fact the result of highly complex underlying processes. After a measurement has been made, Bohm's "hidden" processes act rapidly to randomize the system so that a subsequent measurement appears uncorrelated. If the second measurement were made fast enough, Bohm realized, the system would not have time to randomize completely, and a slight correlation would exist between the two results. In this way Bohm hoped that it would be possible to detect divergences from the predictions of conventional theory.

The problem was that up to now Bohm had been only guessing at the exact details of the underlying processes. Now he and Vigier proposed a particular model and tested out its prediction. The model was that the electron moves in a subquantum fluid. Just as in Brownian motion invisible water molecules buffet tiny pollen grains and cause them to exercise random movements, so too quantum particles are exposed to the causal effects of an invisible underlying fluid. They made calculations based on this model and improved on them over the following years, but the experimental technique was never refined to the point where a crucial test between the two theories could be made.

Conventional quantum theory seemed to Bohm to have arrived at an impasse in terms of the human ability to articulate its meaning. Ordinary language is proving an inadequate tool for discussing the quantum domain, Bohr argued, yet language is the only means by which we can converse about the world. Our human languages are colored by the scale in which we live; they simply do not transpose to the microword. We ask in vain for underlying explanations and pictures of what is happening in the quantum domain. It seemed to Bohm that Bohr had reached a blank wall and there was nowhere left for physics to go, except into some region of ambiguity and mystification.

Bohm, in his hidden variable approach, claimed that it is indeed possible to give a clear picture of quantum processes. His infinity of levels also suggested that physics went on and on, and that the quantum domain was only one stage in the limitless unfolding of nature. This was his vision. Yet his detractors argued that instead of opening out into a new physics, Bohm's hidden variables were now returning to something as inherently mechanistic as a "subquantum fluid." With one hand Bohm appeared to be opening the door to a new, limitless physics, while with the other he was taking physics back to nineteenth-century mechanism.

This was no more his intention than it had been Bohr's to place a limit on the inquiries of physics. Bohm simply wanted to test out a particular model and provide a concrete, albeit simplistic example of how chance events could emerge out of an underlying causal world. In this sense he was touching on something that later became known as chaos theory. Admittedly, as a Marxist Bohm insisted upon a strictly causal interpretation of the world, but for him, this did not mean taking a retrogressive step backward into Newtonian mechanism. Nevertheless, his friends and colleagues must have wondered sometimes at the different directions he seemed to be taking.

Thanks to the stimulation of this active, albeit temporary group, Bohm's former enthusiasm returned. He even began to think about finding a girlfriend. There were "some likely looking girls around here in the Physics Dept and Math," but Brazilian moral attitudes presented a problem. A woman was expected to be a virgin upon marriage, and her family could disown her for having sex with a man to whom she was not married.[13] Bohm would have none of this. When someone suggested to him, "You wouldn't buy a used car if you could get a new car, would you?" he replied that the analogy was overly mechanical. Buying a horse was a better comparison: "You wouldn't take a chance buying a horse that had never been ridden would you?"[14]

One female student caught his eye: "There is one who is rather interesting looking, and who is said to be very far to the left indeed, but I cannot help thinking of her as a little child, which she really resembles. Communication is made more difficult by the fact that these little girls are terrified of big foreign professors. . . . I have too much to do to be interested in forming the characters of little girls. I would rather have one with an already formed character of a kind that satisfies me, (more or less)."[15]

Still, Bohm's interest mounted, even when he discovered that she had become pregnant and was about to marry. She "still retains a child-like wonder and curiosity about the world, while at the same time being mature, intelligent, and cautious enough so that she doesn't let anything be 'put over on her' . . . it is always a cheering sight when I am lecturing to look at her once in a while, but I am afraid that I embarrass her once in a while, because I sometimes feel as if I would like to grab her in my arms and kiss her; and this must reflect to some extent in my expression, even if I try to hide it."[16] If only this girl, he wrote, "hadn't got married so abruptly, damn her she now looks prettier than ever, and seems very happy, while she is also demonstrating more and more intelligence. She is the only girl in Brazil who has thus far had such an effect on me."[17]

Bohm's friend Phil Smith suggested he find a girlfriend in the American consulate. It might also help him to get his passport back. As far as the passport was concerned, Bohm felt it would only work if he went out with Mrs. Shipley, the head of the passport section, "in which case the sacrifice would be too great to be worth it."[18]

After a year in Brazil, Bohm's sexual drive had diminished to the point where "I really cannot remain interested in a woman long enough to get to the point of going to bed, unless there is something more to her than there is to most of the girls I see here. . . . after 2 or 3 days, we would exhaust all topics of conversation, and then contact between us would gradually disappear. . . . I am so constituted that the one thing that is more killing than anything else is 'just live.' "[19]

Bohm could become attracted to a woman only if there was "a

potential for more . . . that is why I don't think I have ever really enjoyed myself after the age of 12 or 13 . . . It is hard to know how this difficulty started, but it probably originated in systematic disappointment in people throughout my childhood and youth—perhaps in the discovery that nobody was really trustworthy and reliable, in spite of appearances. So I have tended to develop a protective habit of not taking people very seriously as individuals. Unfortunately this habit grows, as I discover that in all fields, one feels disenchantment and disillusionment."[20]

Hearing of Bohm's loss of sexual desire, Miriam Yevick prescribed a good time with a "hot babe."[21] To this suggestion, he replied, "I fear that I have practically no desire to have intercourse with a woman who does not already attract me in other ways. I just can't help it, but if you take any old 'hot babe,' the whole business seems to be just a sort of mechanical pumping action, not too interesting in itself, and certainly not worth the long preparation needed before the girl is ready to go to bed." Society's idea of sex was "too simple and mechanical"; it was a myth, created by American advertising and the low position that women occupy in society.[22]

When Miriam argued that a "woman's strongest emotional need is to 'give herself to a man,' " Bohm replied that the desire for subjection is found in both men and women who can become devoted to a strong charismatic leader. Humanity has a longing to devote itself to something higher, he said. When a person discovers an object of desire, a tremendous weight of responsibility falls from his or her shoulders, life is validated, and life's path is certain. The problem, of course, lies in who or what one chooses as the object of worship. Our present society gives every encouragement for women to worship men, at time same time presenting barriers to their independence. The result, he said, is woman's vicarious contact with life though subordination to a man.

In the long run, Bohm argued, this strategy can never work, since no man is worthy of such devotion. But by subjugating herself to a man, a woman establishes a claim on his being that, paradoxically, leads to her controlling him as the object of her will. On the man's part, he may become so conceited as the object of such devotion as to forget his human limitations. Only later, when he is called to account, is he brought back to the human plane in a humiliating way. These were advanced views for a man to hold in the 1950s (but when it eventually came to Bohm's own marriage, practice fell somewhat short of theory).

Bohm noted that a similar form of playacting is present in sexual intercourse. Initially a woman wishes to be dominated symbolically by her partner, ceasing to represent herself but becoming Woman in general. Likewise, her partner represents Man. The sexual act is transformed into a play "dictated by society to represent the relative roles which it has assigned to man and woman." For Bohm, symbolic domination and submission had a dark side, since they furthered oppression by deviating humanity from its natural goal.

In this way the natural desire to serve something greater than ourselves is diverted into illusory sexual outlets. In liberated societies of the future, Bohm believed, sex would be supplemented by other intense experiences, such as dramas, collective singing, dancing, "ceremonies that would mobilize the personality and create a state of mind in which people would feel a unity with the whole of humanity."[23] Bohm himself never entered into a sexual liaison during his time in Brazil—and neither did he experience much in the way of orgiastic singing and dancing! Yet as always, he sought to transcend and go beyond what others willingly accepted as the norm.

After Yevick and Vigier departed, Bohm was struck by the realization that "I depend much more than I thought on such conversation, in order to bring out dormant ideas into a definite form, and to prevent them from continuing to 'sleep' forever, as they tend to do without that stimulation. I find that the 'flow' of ideas is important in this process. A smooth flow causes each idea to lead to another, and I come out with conclusions which I would never have expected at the beginning. This fluent talk is almost as important as the opposition and stimulation of a person who is strongly interested in the same problem, and who understands the background well."[24]

The aftermath of their visit was an empty time for him, and he felt the need for "a sympathetic person to listen to me, to understand to some extent what I am driving at, and to criticize weaknesses in my arguments without attacking the basics; that is, to criticize from the point of view that the basic goal is sound, but that the method may in various aspects be inadequate. . . . my way of thinking is not step by step, but rather through the inter-connection of various aspects to the whole. I never write down any formula before I am already sure of the result on qualitative grounds.

"My way of working (which I am finding effective in science) is to try to avoid exhausting yourself by headlong attacks on insurmountable rock walls. Instead, I like the method of ceaselessly feeling out various aspects of the problem, seeing how things fit together, looking for cracks and weak points. When you find a crack, explore all the surrounding region, and if it looks as if there is a possibility there, you hit that crack with all you've got. Sometimes you are gratified by seeing a very [word deleted] split in a problem that previously seemed impenetrable. But sometimes this means delaying decisive action for a long time. I have found it very profitable to delay an 'all out' attack on the plasma problem for about a year and a half, while I slowly absorbed various aspects, got various ideas, some right and some wrong."[25]

Finally a new intellectual companion arrived in the form of Mario Schönberg, a physicist who had been jailed by the Brazilian authorities for his Communist activities. At first Bohm was irritated by Schönberg's "worship" of Wolfgang Pauli. Even though Schönberg had been jailed for his beliefs, Bohm thought him "the strangest type of communist I ever imagined. He really would pass better as a sort of Jewish businessman." He was a "colloidal character," "only partly sane," "Machiavellian," "very complex, devious and inconsistent," and surrounded by "shady characters."[26] For a time he thought Schönberg was plotting against him, suggesting to others that Bohm's work was "too abstract" and that the university should instead hire a physicist who would make calculations in nuclear physics.

Yet in the long term Schönberg proved an important influence on Bohm. It was he who first pointed Bohm in the direction of the philosopher G.W.F. Hegel, saying that Lenin had suggested that all good Communists read the German philosopher. Hegel's philosophical method stressed the importance of the dialectic as the process whereby a concept or idea, pursued to its limit, engenders its opposite. Thesis and antithesis exist in a paradoxical opposition that is unresolvable at that particular level of discourse. Only by transcending or moving outside the constraints and suppositions of the earlier argument does something creatively new come into existence—the synthesis, or new level of meaning. In turn this new level will eventually be pushed to the limit of its antithesis, and so the dialectical movement engenders a new synthesis.

Hegelian dialectics is a ceaseless process, a movement of thought that is constantly creating and building. It was Hegel's belief that he had discovered the true nature of thought—not thought in terms of particular content but its actual underlying dynamics, and not simply human thought but Thought with a capital T, the World Soul itself, which evolved through a process of dialectical creation and differentiation.

To a committed Marxist, this seemed mere mysticism and needless obfuscation. When Marx and Engels spoke of "putting Hegel on his feet," they meant that they had purged the dialectic of its idealism by applying it to the dynamic processes of matter itself. It was not some mythical World Soul that was in the process of development but matter itself and, because of its materialistic base, human society.

Under Schönberg's guidance, Bohm began to study Hegel's *Logic*. To his surprise, he discovered that rather than setting dialectics on its feet, Marx and Engels had "turned Hegel on his head." Bohm's encounter with Hegel transformed his thinking radically, and over the years to come, he always packed a copy of the *Logic* whenever he traveled. Admittedly his conversion was not immediate, but like the dialectic itself, it developed over time, first in Brazil and later in Israel. Right up to the end of his life, Bohm remained a committed Hegelian who attempted to translate the German philosopher's ideas into physics.

An important example of Hegel's thinking is the dialectical movement from Being to Becoming. Philosophy seeks to begin on solid ground—and what is more secure than the existential fact of Being? Being cannot be denied. Yet as Hegel looked into the concept, he realized that Being must contain within it the notion of non-Being, that which does not exist. Being is different from that which is not. Yet within this very realization, within the boundary between being and absence, is already contained the notion of non-Being.

When pressed to its limit, Being turns into its opposite, non-

Being. How is this paradox—the coexistence of non-Being within Being—to be accommodated in Hegel's systems? Clearly the resolution lies in the very movement itself—that is, in Becoming. Being, Hegel asserts, is in a state of Becoming. The resolution of thesis and antithesis (here, Being and non-Being) is transcendence. Hegel's system thus begins with the static idea, Being, setting itself in motion.

In Brazil Bohm explored how Hegelian oppositions, such as chance and causality, necessity and contingency, could be used in his own work. As he talked with Schönberg, he realized how limited had been his earlier thinking. In his desire to reformulate quantum theory, he had been driven by the need to demonstrate the fundamental nature of causality, and his inspiration had come from Marx. The result had been that he opposed chance and contingency. But now he realized that dialectics could take him much deeper and that his earlier essays and letters must be revised.

Necessity, Bohm saw, is what cannot be otherwise, while contingency is what can be otherwise. Physics itself moves through a dialectical process whereby what appears to be absolute necessity at one level becomes, when viewed from a broader context, the result of contingency. In the early nineteenth century the physicists Robert Boyle and Jacques Charles discovered laws—that is, necessities—that related the pressure, temperature, and volume of a gas. Later, as the existence of molecules was accepted, it was revealed that these necessities were the results of contingency, an astronomical number of random collisions of molecules. Necessity gave way to contingency. But even the apparently random motion of molecules is the result of strict causal laws. Necessity appears again.

Bohm now gave a greater prominence to the role of chance and contingency within his thinking, and sent revised versions of his essays to friends. One of them suggested the collection should be sent to a London publisher, Routledge, and when Bohm did so, he received a favorable reply from its editor, who liked the essays but suggested they be expanded to include more of Bohm's own ideas. Although the final corrections were not made until Bohm arrived in Israel, the bulk of the writing was completed in Brazil. When it appeared in print, Bohm felt that *Causality and Chance in Modern Physics* would be an important contribution to the philosophy of contemporary physics.

Even as he plunged all his energies into the book, he was also complaining about his "frozen state," a "large drop in sexual desire" and "steadily growing loneliness, which literally numbs the emotions." A letter that he wrote along these lines to Lilly Kahler, mother of Hanna Loewy, was shown to Einstein, who (on January 22, 1954) wrote to his friend, commenting on Bohm's feeling of being "closed out and closed in at the same time. . . . What impressed me most was the instability of your belly"—that is, digestive problems—"a matter where I have myself extended experience."[27] The dullness of the intellectual atmosphere "cannot be overcome by the best influence of an intelligent man. The first difficulty could be overcome by a reliable cook, but not the second one." Clearly Bohm must leave São Paulo, but as regards a new position, Einstein had some astute advice to offer: "there is a law in the University-life more so than on the stock exchange: what is offered for sale is falling in price."

At that time Einstein was again waging his old battle against quantum theory. "I am glad that you are deeply immersed seeking an *objective* description of the phenomena," he wrote to Bohm, "and that you feel that the task is much more difficult as you felt hitherto. You should not be depressed by the enormity of the problem. If God has created the world his primary worry was certainly not to make its understanding easy for us. I feel it strongly since fifty years."[28] Several months later, he wrote, "In the last years several attempts have been made to complete quantum theory as you have also attempted. But it seems to me that we are still quite remote from a satisfactory solution of the problem. I myself have tried to approach this goal by generalizing the law of gravitation. But I must confess that I was not able to find a way to explain the atomistic character of nature."[29]

In the nineteenth century the Scottish physicist James Clerk Maxwell had shown that the various phenomena grouped about electricity and magnetism could all be explained by using the concept of a field. Einstein was attracted to the idea of fields to the extent that he believed the whole of physics could be reduced to a set of basic field equations. These field equations would explain the existence of matter, gravity, electricity, and magnetism. But quantum theory subverted this program. Einstein's fields were continuous things, while quantum theory dealt in discontinuities—quanta. In this and several other ways, Einstein's dream appeared incompatible with the quantum theory. Nevertheless he struggled to find a way in which the quantum theory could be absorbed into his basic field approach. The problems he faced were enormous, and after several decades he was almost willing to admit that a radically different approach was needed. "One has to find a possibility to avoid the continuum (together with space and time) altogether. But I have not the slightest ideas what kind of elementary concepts could be used in such a theory."[30]

Bohm believed that Einstein had exhausted the field approach and that something more radical was called for. The problem, as Bohm saw it, was that in his theory of relativity Einstein had created a description of the large-scale universe. Now he wanted to generalize this to include the world of atoms. But as Bohm pointed out, the large-scale world provides very few clues about underlying atomic reality. In fact, the existence of the quantum level had not even manifested itself in physics until the start of the twentieth century. For all intents and purposes, most of physics could continue without ever bothering about atoms, because "the microscopic level is reflected only faintly at the macroscopic level." This, Bohm argued, is the underlying reason why it is so difficult to extrapolate from the large into the small.

Moreover, for Bohm, what physicists take as the quantum level is only the gross manifestation of something even smaller—a subquantum level of continuous and causally determined motion. The basic equation of orthodox quantum theory, Schrödinger's equation, will thus turn out to be only the average of things occurring at a deeper level. But as with the transition from the macro level to the quantum level, the laws at the conventional quantum level are relatively insensitive to the exact details of the new subquantum level. Working from the quantum level, it is extremely difficult to deduce the precise form of underlying laws. It is always possible to make guesses, Bohm told Einstein, but it is difficult to verify which guesses are correct. His only hope was that conventional quantum theory would not apply to very rapid processes. Experiments done in rapid succession would, he hoped, show divergences from the conventional theory and give clues as to what lies at a deeper level.[31]

Einstein, however, retained his deep faith in the essential unity of nature, a unity that had to be reflected in the laws themselves. His theory of relativity might have shown that appearances depend upon the motion of different observers, but the laws underlying these phenomena are not relative; they are universal forms of nature and express the basic invariants of the cosmos. Einstein therefore did not look favorably upon Bohm's vision of a universe of layers. He considered it part and parcel of the hierarchical tradition in physics, in which different levels of reality require different laws. For Einstein, there were no micro and macro laws but only laws "of general rigorous validity, laws that are logically simple . . . and this is the best guide."[32]

Physics must start with simplicity, Einstein believed, and "if nature does not arrange itself that way then we have little hope of understanding it more deeply. . . . If, for example, it is not correct that reality can be described as a continuous field, then all my efforts are futile, even though the constructed laws are of the greatest logical simplicity thinkable." Yet even when he argued with Bohm, Einstein tempered his remarks, adding with characteristic generosity, "This is not an attempt to convince you in any way. I just wanted to show you how I came to my attitude. I was especially strongly impressed with the realization that by using a semi-empirical method, one would never have arrived at the gravitational equations of empty space."[33] Einstein's remarks did not deter Bohm, who continued to look for ways to expand his theory of an infinity of levels.

In April 1954 the Atomic Energy Commission began its hearings into Oppenheimer's security clearance. "I have just heard from a friend," Bohm wrote to Miriam, "that J.O. [Oppenheimer] may be called before a committee soon (of course, he has ways of getting out of it perhaps)." Oppenheimer could easily name names, Bohm realized. (During the war Oppenheimer had informed the authorities that Bohm was potentially "dangerous.") To Miriam, he referred to "the great man himself, whose face once appeared with infinite sadness on the cover of Time Magazine. Perhaps he will really have reason to be sad too. As I once said he looked like 'Jesus Christ' in the picture. I think a better image would be a linear combination of J.C. and Judas, or of Judas trying to look like J.C. An interesting case of mistaken identification, don't you think."[34]

Throughout Bohm's life he experienced a series of betrayals at the hands of men he admired and saw as father figures. As a child he had felt rejected by his father and, in turn, had turned his back on all that his father stood for. For a time Oppenheimer had been the wise and understanding man who inspired love in those around him. It was in the atmosphere he created at Berkeley that Bohm had discovered Marxism and began his first creative research. Now the man he had once admired was almost eager to betray his students and colleagues.

Years later, in 1966, Bohm wrote to Oppenheimer: "I myself have begun to feel that nothing is to be gained by raking over all those dead ashes again and again. . . . If Mr. Stern's book [Philip Stern was writing a book on the Oppenheimer case] were able to help us learn from our mistakes, it might be useful. Whether or not it will in fact do so is not at all clear to me at present. . . . I can understand that your dilemma was a particularly difficult one. Only you can assess the way in which you were responsible for what happened, not because you made a simple mistake, but because your mistake flowed out of contradictory basic principles to which you tacitly (and perhaps unconsciously) adhere."[35]

In this same period another member of the Berkeley group, Joe Weinberg, was indicted for espionage, having been identified as the notorious "Scientist X" who allegedly passed atomic secrets to the Soviet Union. "I think that poor Weinberg is a dead duck, in the atmosphere that exists today," Bohm wrote. "Heaven knows how far they will go." During Weinberg's trial no evidence was presented by the prosecution, and the physicist was acquitted.[36]

It seemed to Bohm that a poisonous atmosphere of suspicion and accusation prevailed in the United States as never before. He seriously believed that concentration camps would be established for the incarceration of Communists and other dissidents. He was particularly concerned for the fate of Melba Phillips, "a courageous woman" and one of the few "sparks" left in the United States.[37] Like Rossi Lomanitz, she was finding it difficult to obtain an academic position, but Bohm saw even greater danger ahead and pointed out that it was her "duty to stay alive." As to his regular exchange of letters with Miriam, he told her: "Perhaps it would be wise for a time to call a halt to political discussions, especially since letters may be read in this time of movement toward Fascism."[38]

The political environment in Brazil was equally depressing;

maybe it would be better in Europe if only he could get there. Friends in Israel now put him in contact with one of Einstein's former assistants, Nathan Rosen, at the Technion in Tel Aviv. But the problem remained that of Bohm's lack of a passport. For a time he considered making an appeal to the American consul, but then he worried that people in the consulate would create trouble for him with the Brazilian authorities once they learned of his intention to travel. His only hope, he thought, was to take out Brazilian citizenship. He wrote to Einstein for advice.

Einstein agreed that Brazilian citizenship was the best solution and said he would do everything he could to help Bohm obtain a position in Israel. Finally Bohm received a firm offer from Rosen. Then at the very last minute, and despite years of complaints, Bohm decided to stay on in Brazil. At last he had found a place where the food did not make him ill, his work was going well on several different fronts, and Schönberg was knowledgeable about the mathematics he needed. A few years earlier he had compared the task facing him as requiring the abilities of "a Newton, Einstein, Schrödinger and Dirac all rolled into one." Now he believed that his theory could do everything that conventional quantum theory had achieved and far more. He was going to explain that mysterious connection whereby elementary particles with integral spins obey Bose-Einstein statistics while those with fractional spins obey Fermi-Dirac statistics; he would also describe the creation and annihilation of elementary particles in terms of a subquantum ether.

By the late summer of 1954, Bohm felt himself "intellectually sharper than ever."[39] At last physicists would surely be convinced by his theory. A few weeks later, however, he fell into depression. "I have reached new impasses in my work and I seem also to be losing my strength. Maybe I am getting old, who knows. I seem to have no sexual feelings or desire left whatever; in fact the whole region between the legs feels as if it were quite empty. . . . I just don't have as much energy and adventurousness as I used to have. I don't get such a kick out of new ideas."[40]

By November, his depression was "steadily deepening"; he was "really fed up with Brazil, with no one to talk to and nothing to do at all in my spare time. . . . I am approaching a time in my life when this depression and isolation is becoming more dangerous to my health." He was genuinely concerned that he no longer had sufficient strength to pull himself out of his state. "I am beginning to feel cold inside, a coldness based on a steady growing sense that I will never have the kind of life that I have wanted, a coldness that seems to foreshadow that of death."[41]

His only hope was escape, and within a matter of weeks he accepted the Israeli offer and became a Brazilian citizen. Bohm left the country in mid-January 1955, stopping off in Argentina to talk with philosopher of science Mario Bunge. From there he flew to Europe, spending a week in London and three weeks in Paris so that he could work with Vigier and de Broglie.[42]

Miriam had now given birth to a son, David. From Paris, Bohm wrote to congratulate her and promised to discharge his duties as "god-father, rich uncle . . . teacher of philosophy and physics, and perhaps even a little politics." He did not, however, approve of his namesake being bottle-fed. He had intended to buy a gift for "young Davey," he said, but then "Jean-Pierre [Vigier] would start by discussing physics and before we knew it, it would be too late to do anything."[43] In the spring of 1955 he arrived at the Technion in Israel.

Throughout his Brazilian exile Bohm's mood had often been dark and depressed. His letters spoke of the death of feeling within his body and periods in which enthusiasm for his work totally vanished. Yet it was in Brazil that he had a vision of the infinity of nature; in Brazil that he had encountered Hegel and written his *Causality and Chance in Modern Physics*. He, too, gave a gift to Brazilian science: When he had arrived, the physics department had been in a shambles and the students dispirited. By the time he left, as one of that department's physicists, Elsa Faher, put it, Bohm had created an animated and overcharged group of physicists.

Chapter 10
Israel: The World Falls Apart

SHORTLY AFTER HIS ARRIVAL in Israel, Bohm wrote to Miriam, "I have begun to go around with a girl, named Sarah Woolfson from England. . . . I like her quite a bit. Perhaps she is something like Hanna, but with more energy and courage."[1]

Sarah Woolfson, the daughter of a doctor from Reading, England, had worked as a physiotherapist until, during the 1948 emergency, she and her brother traveled to Israel as hospital volunteers. Sarah—or Saral, to use the name by which she was always known—enjoyed her new life so much that she decided to stay on. A year later she was joined by her parents, Dr. Woolfson taking charge of physical medicine in Haifa.[2]

Saral loved music. While she was in Jerusalem, she lived in the home of the cellist Thelma Yellin. When traveling, Yellin would ask Saral to make her home available to musicians for practice in her living room. Among the many musicians who visited was Isaac Stern, who would spend many hours at the house rehearsing before a concert. Saral also had a French horn player as a boyfriend, who had graduated from the Juilliard School.[3]

From Jerusalem Saral moved to Haifa, and in the spring of 1955, at thirty-two, she was staying with her parents while looking for new accommodations. Across the road lived a physicist named Mort Chwalow from the Technion. Saral enjoyed his parties, and on one occasion she noticed a young man sitting in the corner looking lost. Having no idea of his scientific reputation, she was suddenly struck by the intuition that she would spend the rest of her life with him.

Saral introduced herself, and Bohm told her that he had arrived in

Israel a week or so earlier. She explained her work with children who had been crippled in a recent polio epidemic. At their little farm on the kibbutz, it would soon be time for the spring festival, where the children would present the first fruits of their produce. It would be an attractive ceremony, she said, with the children carrying a lamb, chicken, and fruit in a basket. When Bohm seemed interested, she invited him to join her.

Her friends were surprised at Saral's spontaneous approach. Bohm was such a famous scientist, they told her, that he was unlikely to go. Nevertheless, Bohm was ready when she came to pick him up, and during the festival they walked to the sea, talked, and learned more about each other.[4]

Soon Bohm was visiting Saral's apartment, where they would lie together, holding each other, listening to music or simply talking. Bohm's own apartment was located at the top of Mount Carmel, and each day he chose one of several paths for his hour's walk down the mountain to the Technion. Somehow Saral always knew which path he would take and was there to meet him. Their relationship developed rapidly, and soon they were living together.

With her cheerful, optimistic, and outgoing personality, Saral Bohm is the sort of woman who enjoys social relationships, inviting people into her home and providing a meal. She is comfortable with the physicality of life, and had worked with the human body as a physiotherapist and learned to sculpt in clay and plaster. In many ways she balanced and complemented Bohm's overly intellectual approach to life and his emotional reserve and distance. In Bohm's mind she may have evoked memories of the Polish and Irish mothers of his childhood and the warm, hospitable homes they created.

Saral offered him stability and relief from the loneliness that had been his constant companion in Brazil. On a practical level she made his life easier by cooking good simple food, so that his stomach problems quickly cleared up.* It goes without saying that his loss of physical desire was also rapidly cured.

* Bohm always paid close attention to his diet, but he did not always exercise common sense. When we worked together in upstate New York, he would sometimes take me to a kosher restaurant in nearby Mount Kisco. There he and Saral ate well and then, against their better judgment, invariably indulged in a very rich dessert. Next day, David would greet me sluggishly, complaining that he hadn't slept and for some reason felt unwell.

Later in life Bohm abstained from alcohol, his only stimulant being coffee. But in Israel he enjoyed drinking wine at meals. Saral noticed that, after a few drinks, he became jolly. Alcohol was good for him, she concluded. At one party a fruit cup heavily laced with brandy was served, she later remembered. It was a hot night, and unable to taste the alcohol, Bohm drank it like lemonade. As the evening progressed, he became ever drunker, giggling, laughing, and looking extremely happy.

Bohm's first contact with Saral lasted for only a few weeks, as he was dividing his time between the Technion and Europe. In July he was off to Miriam's native Holland, where she and George had taken a summer cottage at Loosdrecht. There George intended to spend several intense weeks exploring the connections between Bohm's earlier plasma work and his new theory that the electron is a process in an underlying ether.[5]

However, George arrived with a copy of Hegel in his suitcase, which subverted his whole plan, since Bohm pounced on the book and began to talk philosophy. Miriam was preoccupied with fourteen-month-old Davey, yet even in the midst of a domestic crisis, Bohm followed her around, going on and on about dialectics.[6]

When they were able to take a break from the German philosopher, they all rented bicycles and went on trips together. Relaxing with his old friends and playing with his godson gave a boost to Bohm's spirits. Vacation snapshots show a smiling face and a relaxed manner. But suddenly, in the midst of it all, he felt dizzy and complained of heart palpitations. The Yevicks called a doctor, who admitted Bohm to the hospital. The disorder was never diagnosed, but bed rest certainly helped. The Yevicks wondered if the whole thing had been a reaction to his years of tension and agitation.*

Although Bohm was a guest of the Yevicks, the question of his sharing their living costs arose. Miriam suggested that since Bohm enjoyed drinking coffee so much, he should pay for it. Bohm was accustomed to drinking a great deal of coffee in Brazil, but the price of

* Bohm later remembered that during the vacation his breath and urine had smelled strongly of acetone and his diet had included a large number of sweet Dutch cookies. His boyhood addiction to sugar had not abated, all of which points to a borderline diabetic condition.[7]

coffee in Europe was much steeper. When the bill was presented, it generated a degree of friction. Finally Bohm agreed to pay, but he explained that he was worried about money and, not knowing what the future held, was forced to count every penny.[8]

While Bohm was generous in so many other ways, Miriam noticed, he was always extremely careful about money. It seemed to her an aspect of his general anxiety; part and parcel of hypochondria, fussiness, and fears. Although Bohm despised his father's concern with profit, finances came to occupy an important position in his own mind, and money worries dogged him throughout his life. Decades later his colleague at Birkbeck College, Basil Hiley, recalled that during departmental meetings Bohm would sit in obvious boredom, doodling away on a pad of paper. Only when a passing reference was made to examination fees did Bohm grow alert, interjecting, "How much did you say? How much?"[9]

Holland was relaxing for Bohm, but George was frustrated that his friend talked endlessly about philosophy, never coming around to scientific questions. To make matters worse, Jean-Paul Vigier arrived from Paris to stay at the nearby seaside resort of Zandvoort. Bohm immediately began commuting to Zandvoort to resume their work on the causal interpretation. This was the last straw for George, who accused Bohm of being more interested in Vigier than in his old friends. He had come all the way from the United States to do physics with Bohm, he said, but all he ever heard about was Hegel, Hegel, Hegel.* At this Bohm hit back: "If that's how you look on it then you should have paid me. If I'm here to do you a service then you should have made it perfectly clear." George's forgiving nature could not be suppressed for long, and soon they were again the best of friends. But nothing in the way of scientific collaboration materialized during that visit.[10]

* I can well commiserate with George Yevick. There were times in the 1980s when I would drive down to New York from Ottawa to work with Bohm on our book, *Science, Order and Creativity*. Although we both knew time was limited, Bohm would go on, day after day, about some completely different topic that currently engaged his interest. When he got his teeth into an idea, it was useless to try to distract him. The best one could do was go along for the ride and hope that his latest obsession would eventually exhaust itself. His monologues could be unceasing. His friends of later years, Mary and Alex Cadogan, once made a pact to remain totally silent in Bohm's presence, just to see how long he would continue. He talked for around an hour, paused for a moment, then began again.

The meeting in Holland revived Miriam's romantic interest in Bohm, however, and she wondered if it was still possible that they live together. But Bohm never made that final act of commitment and ask her to leave her husband. Maybe he was truly indecisive, or maybe he had lost interest. George was still his friend, and after all, he had begun a relationship with Saral Woolfson back in Israel.

After he took his leave of the Yevicks in Holland, Bohm traveled to Paris, where he intended to continue his work with Vigier. But here the tables were turned, for French politics were particularly lively at the time, and Vigier spent most of his time at Communist party meetings and demonstrations. In this period Vigier had little time for physics and none at all for Hegelian dialectics.

While he was in Europe, Bohm kept his eyes open for possible university positions, and he took a side trip to England to give seminars. His talk at University College, London, proved particularly significant. In the audience was a bright young physicist, John Bell, who was also struggling to understand the role of causality in quantum theory. For Bell, Bohm's hidden variable papers came as a revelation, seeing in them "the impossible being done." He was equally struck by Bohm's lecture, and from that point on he was a staunch supporter of his work.

Inspired by Bohm's treatment of the Einstein-Podolsky-Rosen paradox, Bell went on to develop his own Bell's theorem, which put an end to Einstein's belief that "independent elements of reality" exist in a quantum world. Quantum mechanics, Bell showed, is inherently nonlocal and transcends the limitations of classical space and time.

Einstein, during his arguments with Niels Bohr, had clung to "independent elements of reality," the idea that quantum systems, far from any disturbance or observer, are well-defined and real. To this end he proposed a thought experiment in which one of a pair of correlated but widely separated particles is observed and its parameters noted. Thanks to the fundamental conservation laws of nature, it is then possible to make inferences about the properties of the second particle. In view of the great distance between them, Einstein argued, the fact of observing one of the particles can have no physical effect on the other. The undisturbed part of the system must logically have an independent reality. But if this is true, then it is indeed possible to speak of "independent elements of reality."

Bohr's counterattack had been subtle, at times to the point of obscurity. In essence he argued that even when well separated, the quantum system is nevertheless an unanalyzable whole. Einstein's position, he said, is invalid—there are no "independent elements of reality" in the quantum world.

In *Quantum Theory* Bohm had set out to refine Einstein's thought experiment. As before, a composite particle splits in two, but this time it is the spins of the components that correlate. Because spin is a discrete quantum property, the nature of the paradox is made more clear. Still, even in this case the paradox remained a purely hypothetical thought experiment.

John Bell took the next step, not so much by extending Bohm's example in *Quantum Theory* but by returning to Bohm's 1952 papers on hidden variables. For Bell, these papers presented a totally consistent alternative to conventional quantum theory. Moreover, the hidden variable theory was both causal and nonlocal; nonlocal in the sense that distant parts of the experimental apparatus affect the outcome of a quantum observation. This nonlocality, which goes beyond anything in classical mechanics, had also been implicit in Bohr's discussions of quantum wholeness. Only now, thanks to Bohm's alternative interpretation, was it made explicit.

After reading Bohm, Bell began to wonder if nonlocality is an essential feature of all quantum mechanical accounts. Certainly other physicists had attempted their own versions of quantum theory, based, for example, on making variations and additions to classical mechanics. But as far as Bell could see, all of these theories remained essentially local. To take a crude example, a theory in which the electron is pushed around mechanically by tiny subquantum particles, like a particle undergoing Brownian motion, is a purely local theory.

Bell went on to establish a formal criterion for locality within all classical theories. This involved establishing mathematical limits on the degree of correlation between two particles. Bell's contribution was to think up a set of experimentally testable conditions that could be created in the laboratory. Bohm had earlier shown how the EPR paradox could be expressed in terms of correlated spins. Bell now generalized this insight into an experimentally testable situation, making use of the fewest possible assumptions. He established rigorously that quantum (and Bohmian) correlations exceed all correlations that can be established within any possible classical theory—that is, one involving fields of interaction or the exchange of physical particles.

Soon after Bell's theorem was published, a number of experimental tests, each one more refined and each one designed to overcome possible objections, confirmed the essential nonlocality of the quantum world. Once and for all, Bell had established that correlations exist at the quantum domain, correlations that can never be explained in terms of local, classical theories. Even if, in decades or centuries to come, quantum theory is overthrown, the theory that replaces it must reflect the essential truth of Bell's theorem—that is, the nonlocality of subatomic nature. Several years after the first confirmation of Bell's correlations, there was talk of Bohm and Bell sharing a Nobel prize, along with the experimentalists involved.

At the end of the summer, Bohm was back in Israel, where his relationship with Saral Woolfson deepened. His life, as usual, was surrounded by the familiar miasma of complaints. The heat bothered him, everything was badly organized, and the society was corrupt. To make matters worse, his colleagues at the Technion, David Fox and Paul Ziszal, were planning to return to the United States. Bohm's only hope lay in Europe, and so he began to think seriously about offers that he had received from Maurice Pryce at Bristol University and from Louis de Broglie in Paris. Inevitably he agonized about his future, to the point where he could no longer concentrate on his work.

Although Bohm did not find the Technion particularly stimulating, he did make contact with two outstanding students, Yakir Aharonov and Gideon Carmi, who were to become his future collaborators. Carmi excelled in several fields. As well as having a talent for physics, he was a good pianist and painter. He had also studied with Moshe Feldenkrais the system of subtle body movements used in therapy. Carmi explained to Bohm that he believed in a deep connection between physics, consciousness, and these subtle, minimal movements.

Yakir Aharonov was considered something of a maverick in the physics department because of his obsession with a putative new effect in physics. Who better for him to talk to than the arch-outsider

David Bohm? The two students, Carmi and Aharonov, struck up a friendship with Bohm and later followed him to Europe.

Bohm's most important relationship, however, remained that with Saral Woolfson. In those early days they had great joy in their relationship. Bohm would hold Saral, kiss her warmly, and touch her hair. Music was important to Bohm, particularly chamber music, but theater was less so since he found it difficult to follow the actor's voices. He enjoyed films, especially a psychological thriller, a nature film, or the English humor of the Ealing comedies.

Bohm's own sense of humor, and of the absurd, was keen. Later in life, when he became a regular visitor to the United States, he was highly amused at the headlines on tabloids like the *National Enquirer*. He enjoyed scanning through them at the supermarket checkout while Saral dealt with the groceries—"Eighty-year-old woman gives birth to Elvis's love child," "Egyptian pyramids found on Mars," and the like.

He enjoyed jokes that involve wordplay, puns, and verbal paradoxes. He would often return to the concept of "nothing" or "no one." "When Saral came back from the shops today she saw no one," he would say, "We saw him yesterday, do you remember? Maybe 'no one' will come to the house this afternoon."

Bohm had great pleasure in visiting Saral's relatives in Israel. But inevitably the children would take him aside and ask him for help with their homework. Soon he would be sitting with them and working out their mathematics problems.

Above all, Bohm loved to be with and talk to Saral. On many evenings they would sit together until it was cool enough to go outside and walk. She later remembered one evening particularly vividly. It was a clear night with showers of shooting stars and a moon so bright it was possible to read a book by its light. They started high up, on the top of Mount Carmel, and walked along the wadi, or dried riverbed, to the sea. It was a five-hour walk though pine trees, live oaks, and the scent of wild flowers, sagebrush, and pine needles. All around them the grasshoppers chirped—for once they were content to walk in silence. On other occasions the night was so magical, they would lie together under the pines. During such moments Saral felt a great openness and affection for Bohm.

The couple also liked to visit the Haifa concert hall to listen to chamber music, particularly Mozart and the late Beethoven quartets. Afterward they walked home hand in hand "singing" the music they had heard. There was something so sublime about Mozart's music, Bohm told her, that it transported him into the realm of clarity and coherence. Looking back on her husband's life after his death, Saral was struck by the fact that Mozart had produced such music in a life of turmoil, and she asked herself the question that Peter Shaffer had posed in his play (and later the film) *Amadeus*—where did the music come from? From his earliest days Bohm had experienced pain, disappointment, and even what could be called betrayal. Yet always he expressed joy and excitement at the possibility of going beyond, of discovering coherence and wholeness. So much of Bohm's work spoke directly to others and enabled them to achieve a sense of wholeness, yet at the same time, so much of what he did was born out of his own suffering.

But in those early days with Saral, that paradoxically painful element in Bohm's life seemed far away. They were in love and simply enjoyed being together. As they walked at night, Saral would ask, "What is relativity?" or "What is quantum theory?" and Bohm would patiently explain the concepts in a clear and simple way. Once he told her that it was impossible to transform a left-handed glove into a right-handed glove without changing every atom. That idea so captured her imagination that she created a sculpture based on it. If only science could discover the true nature of matter, he told her, then people would be able to understand consciousness and society. Bohm reminisced about Einstein and how, when he talked, the German physicist seemed to grow, to become so big and full of energy. It was clear to Saral that Einstein thought a great deal of Bohm. Some years later, when Lilly Kahler visited the Bohms in London, she told Saral that Einstein had looked upon Bohm as his "intellectual son."

Bohm loved to talk philosophy, about the nature of thought, and his overwhelming passion for Hegel. When he explained the dialectic, Saral was able to capture the beauty of the idea. At other times he seemed so content that they simply walked in silence. Yet even in their closest moments Saral was aware of Bohm's reticence and the difficulties he experienced when she asked him about his inner feelings.[11]

In his previous relationships with women, Bohm had exhibited a degree of dependence and evoked in them the desire to protect him. Saral saw this pattern in a broader context than that of a mere neurotic need on his part. Bohm was a unique individual, an exceptional mind, able to pursue abstract flights of thought to a far greater degree than other people could. But it is difficult to live in high abstraction without losing one's equilibrium. Bohm needed an anchor in life, someone who would give him the stability and, to the extent that it was possible, provide him with a normal life.

Like many other creative people, Bohm was prone to recurring periods of depression. Having extremely high expectations meant that he was constantly in danger of being let down. Saral tried to support him during these times; she felt that, without her help, there would have been occasions when he swung even deeper into a blackness that could have cost him his life.

Keeping Bohm's feet on the ground meant never allowing him to become inflated with the feeling that he was in some way privileged. This feeling was not always easy to limit since, when he talked to a small audience, he could be a powerful, charismatic figure. As a result, some people made him the object of their projections. Generally Bohm could keep a sense of reality quite well, for his sense of humor gave him a keen nose for the absurd, and a witty remark can always deflate a potentially embarrassing situation. Yet running counter to this sense was his inability to say no when people demanded his time. Either he would put up with their unwelcome intrusion or else he would mumble something to the effect of "Yes, well, erm, maybe. Perhaps tomorrow." Rather than refuse an invitation outright, he preferred to evade and postpone. This character defect caused genuine confusion among his colleagues and associates. When people were inviting him to a conference, they were never sure if Bohm's hesitancy meant he had reservation about the nature of the meeting or if he was simply being vague.

Bohm needed time for his thinking, and it fell to Saral to screen his telephone calls and requests for appointments. This role led to a degree of hostility toward her. Those who, in their self-importance, believed they had a right to Bohm's time, or that he would naturally be fascinated by what they had to say, became offended when she deferred their calls. Saral also had to face the antipathy of those women who believed, or fantasized, that they could provide Bohm with the ideal relationship and environment for his work. It should be added that right up to the end of his life, Bohm could exert a considerable charm over women. As a young man he was handsome, but his greatest attraction lay in the passion of his mind. Talking to Bohm was an overwhelming experience, as energy, insight, and intelligence poured out of him. Women who bathed themselves in this passionate illumination could easily fall in love with him. Bohm, for his part, was probably quite unaware of the effect he was producing.

In short, living with David Bohm proved to be far from easy. Saral foresaw this, yet she believed the relationship was inevitable. Still, looking back after her husband's death, she sometimes wondered why she had committed herself to such a hard life. Her conclusion was that her relationship with Bohm was a responsibility she simply had had to take on. In the last analysis, their relationship was based on love, and if David was not always able to express open affection with ease, then at least, just before he died, he told her how much he loved her and how important their relationship had been to him.

The ties deepened, and before the year was out, their affair moved toward its next phase, marriage. Bohm asked Saral to come with him to Paris. While there was no problem about their living together unmarried in Israel, a European trip would make the relationship explicit, and Saral knew this would embarrass her English relatives. The best solution would be to regularize their condition, so if Bohm wanted her to travel with him, then he had to marry her first.

Saral talked this over with her friend and colleague Dr. Weiss, a psychoanalyst who had been one of Freud's last students. Weiss liked Bohm, but seeing the dangers in marriage to him, he strongly advised her against it. He explained what such a relationship would entail and how difficult it would be for David Bohm to be "tied down."[12]

Weiss was correct, and in the weeks leading up to their wedding, Bohm vacillated and worried about whether he was doing the right thing. It was not a good time for him to take such a step, he suggested; maybe they should wait. To make matters worse, in the three years since Stalin had died, rumors had increasingly circulated about the repressiveness of his rule and about the wretched social and political conditions in the Soviet Union. Bohm tried to deny these stories as distortions of the yellow press or as weak-minded reactions of liberal "lefties," or else he tried to justify them as the unfortunate but necessary breaking of a few eggs to make the Soviet omelet. While many of Saral's friends were politically to the left, even she was surprised that such an intelligent mind as Bohm's could accept Communism so uncritically. The last straw came in February 1956 when, during his address to the Twentieth All-Union Congress of the Soviet Communist Party, Nikita Khrushchev denounced Stalin, listing his crimes against humanity and his terrible history of economic mismanagement. Yet even then Bohm seemed able to close his mind to what was appearing in the papers and to retain his dream of an ideal Marxist society.

Although Bohm's anxiety increased as the wedding date approached, he did not back down. But now he encountered a new difficulty. There were no civil marriages in Israel, but for the religious ceremony to take place, both parties had to produce witnesses who were Jews and could testify that there were no impediments to the marriage. Bohm did have a friend in Israel who had previously known him in Brazil, but he wasn't a Jew; none of his Jewish acquaintances has known him long enough to swear that he was not previously married. Finally bureaucracy was satisfied, and on March 14, 1956, the Bohms were married in Haifa.

Immediately afterward, David entered into an acute and agitated depression. He felt trapped, unable to move; the whole thing had been a terrible mistake. His anguish became so great that Saral had to reassure him that both of them were totally free to walk out of the relationship anytime they wished and that in Israel, for a man, divorce was a particularly simple matter. But her suggestion that they need not live together raised an even greater turmoil in Bohm's mind. On the one hand he panicked, feeling trapped, but at the same time he realized that he could no longer be without Saral.

So intense was Bohm's depression that Weiss feared he would commit suicide and advised Saral to watch him day and night. But soon the worst of the crisis was over. Weiss told her that Bohm was no longer in danger and they were free to travel to Europe.

The crisis of marriage eventually gave way to quotidian married life, where more longstanding psychological problems arose. From his early childhood with an unstable mother, Bohm had found it painful to demonstrate his inner feelings, and he felt truly safe only in the company of animals and children. His feelings were always armored against the world. Saral was forced to accept that sharing his ideas with her was an important way by which he could demonstrate his affection for her.

Many people who encountered Bohm casually found him highly reserved and distant. Others, who got to know him better and saw beyond this, sensed the warmth that accompanied his energy and intensity. On a personal note, I was aware of the friendship Bohm offered me, yet I realized that I must always remain sensitive to the emotional distance he needed in order to feel comfortable. Following the completion of the final chapter of our book *Science, Order and Creativity,* David walked me to the Edgware tube station—I was about to fly back to Canada from London. At the entrance we said our good-byes, and he clasped my hand to shake it. Then he held on to it, pressing it for a moment longer. He looked into my face, and I experienced such an intense emanation of emotion from him that it moved me almost to tears.

In the light of his mother's violent and manic reactions and his father's rages at her inability to look after the household, Bohm found it difficult to deal with anger in others. Strong emotion made him uneasy to the point where he felt he had to retreat.* In the first months of their marriage, this aspect of his personality became a serious problem. Each time Saral tried to express her normal anger, David would become frightened, later explaining that it was because of his mother. It goes without saying that no wife wishes to be told that she reminds her husband of his mother! But Saral trod carefully, and the relationship stabilized.

With Bohm's mental equilibrium restored after the wedding, he and Saral traveled to Europe to work at the Institute Henri Poincaré with Vigier and de Broglie. Over the next years these excursions became a regular summer break for the Bohms, who stayed at the

* With his wife's increasing inability to cope with daily life, Samuel Bohm had been forced to look after his two sons. This included giving them breakfast each morning. In Saral Bohm's opinion, the monotonous diet he provided was the source of Bohm's lifelong antipathy toward cornflakes!

Hôtel des Beaux Arts, in the rue des Beaux Arts, or at the Hôtel d'Anglais on the rue Jacob. On one of these visits they were disturbed by their neighbor on the rue Jacob—an Irishman given to bouts of heavy drinking, which ended in his vomiting on the stairs late at night. One day while David was at work, this man "chatted up" Saral almost to the point of propositioning her, until she became quite angry. Later she discovered that her neighbor was the playwright Brendan Behan.[13]

Bohm spent his days with Vigier and de Broglie, leaving Saral free to pursue her art classes and lessons. Under the influence of an instructor who had seen some of her life drawings, she began to paint in addition to continuing with her sculpture.

In the evenings Saral and Bohm enjoyed eating out in local restaurants. One night, as they were walking home, Saral realized that Bohm was wearing the wrong overcoat. She told him to return to the restaurant, but he walked on, saying that things were perfectly all right because the owner of the coat he had absentmindedly taken would be able to go home in Bohm's coat. Characteristically, Saral noticed, Bohm got the worst of the bargain, since his own coat was new and more expensive; later he went back and retrieved it.

Entering into marriage had precipitated a psychological crisis for Bohm. Now, when Saral left him in Paris to visit her relations in England, he was plunged into another attack of anxiety. The seeds of this latest crisis lay in Khrushchev's denunciation of Stalin. The Paris Communists were in turmoil at the news; some resigned from the party but others, like Vigier, argued that Stalin's personal corruption did not logically negate the validity of Marx and Lenin. To make matters worse, Vigier and his wife played out this schism on a personal level, and Bohm found it difficult not to become embroiled in their constant arguments. At last he was being forced to face the full implications of reports that he had attempted to deny as exaggerations, Western propaganda, or deliberate lies.

During his exile in Brazil, Bohm had kept before him, as a talisman, that dream of the Russian experiment. Admittedly its realization had difficulties, but if only it could be allowed to continue beyond its teething problems, then a better, freer, and more rational human society would emerge. Now he was forced to admit that Stalin's rule in the Soviet Union had been even more ruthless than that of the czars. Had he been deliberately naive all those years, Bohm asked himself, closing his ears to what everyone else had been telling him for so long? With Saral away, Bohm walked the streets of Paris for hours. His entire ideological world was collapsing around him, leaving him nowhere to stand. There was a vacuum at the center of his life, and as every scientist knows, a vacuum is something nature abhors. From somewhere that vacuum had to be filled.

Saral learned of the seriousness of Bohm's state when she telephoned him from England. At once she urged him to join her. But as soon as he arrived in England, a new political storm erupted. On June 24 Colonel Gamal Abdel Nasser had been elected president of Egypt, and a month later he nationalized the Suez Canal. The result was military retaliation on the part of France and England. Their action was based on a complicated but dubious web of mutual defense and nonaggression treaties, which also involved Israel. International conditions became so unstable that Bohm believed that Israel was about to be invaded, and he worried about returning to the Technion. In the end, the Anglo-French intervention was tempered by international criticism, and the crisis petered out. The European action did, however, create a moral void in which another military operation could take place—the Soviet invasion of Hungary. This move was yet another blow for Bohm. At this point many European Communists left the party.

Most of us are concerned to some extent about international politics, but Bohm's concern was qualitatively different. He took the world's problems onto his own shoulders and agonized about what should be done. Corruption, political mistakes, and military actions, he believed, were all evidence of deep errors in human thought and society. And since Bohm believed in the wholeness of the world and consciousness, these errors were also enfolded within his own thinking. In some sense he believed he shared responsibility for what had happened. His own efforts had not been sufficiently vigorous, his arguments not perfectly clear and persuasive.

Again and again Bohm told his friends that if only human beings could understand the underlying rationality of the physical world, they would begin to lead their own lives in a rational way. This was one of the most important reasons why the Copenhagen interpretation of quantum theory, with its denial of causality and its inadequate metaphysics, had to be combated. Bohm's own causal interpretation was the answer, he was certain, a demonstration that pure rationality exists right down to the level of the atom. But his arguments had not been heard, his appeal to lucidity ignored. And, as such, he felt responsible for everything that was disintegrating around him.

That year, 1956, shook Bohm's beliefs to their foundation. No longer could he have faith in the absolute power of science, logic, and human knowledge. Human nature was inherently flawed, he realized, and no appeal to science, and no political theory like Marxism, could cure it. Probably without realizing it at the time, Bohm had begun the search for an understanding that lay far deeper.

Despite the pain of political upheaval, life was not totally bleak. Upon his return to Israel, Bohm made contact with the philosopher Mashulan Groll. Groll, an active proponent of the kibbutz movement, was an authority on Hegel's philosophy. Bohm attended Groll's seminars in Tel Aviv on the nature of the dialectic, then visited him in Jerusalem. There the two men spent a day attempting to talk about Hegel. Groll conversed in German and Hebrew and, while he spoke no English, he could understand a little of the language. Bohm, for his part, spoke no Hebrew but could read some German and understood a little of it when spoken. As Groll talked, Bohm sat with the English translation of Hegel, and when concepts became particularly difficult, Saral translated between Hebrew and English.

Bohm quickly realized that Groll had a deep insight into Hegel. The philosopher's writings on Essence and Actuality, he learned, were of key importance. In the years since Bohm's conversion to Hegel in Brazil, his ideas had matured. He had previously conceived of the dialectic in Marxist terms, and despite the idealism of Hegel's logic, he still believed that the essence of reality lay in matter itself. Nothing existed except matter, and human behavior arose out of material processes. It should be added that by *matter* Bohm meant something that went far beyond naive materialism for, to him, matter was inexhaustible and could never be contained within fixed physical laws.

His Marxist faith shaken by the political events of 1956, Bohm moved, as he talked with Groll, toward a form of idealism, giving a fundamental role in the cosmos to thought. Marx and Engels had profoundly distorted Hegel's basic position, he saw. The dialectic, as he had realized in Brazil, was not primarily concerned with matter but was in fact the basic movement of thought. And not simply thought in terms of the particular contents of consciousness—particular thoughts about the world—but rather the entire process by which thought operates.

As a boy crossing the river, Bohm had seen the world as a flowing movement. Now he realized that Hegel had revealed the nature of this most basic movement of reality. Thought is movement; yet thought also attempts to hold fast to itself and seeks security. It does so by entering more deeply into a particular thought. But each particular thought is limited, and by entering into it, thought must eventually reach contradiction. The basic movement that occurs when thought attempts to hold fast is to enter a region where a particular thought eventually enters into this contradiction. Contradiction has the effect of taking thought from the particular into a region beyond. Therefore in the very act of arresting thought, of seeking certainty, a new movement emerges.

This basic movement of the universe of thought was very different from the struggle Bohm had earlier pictured within Marxist dialectic. The whole movement of consciousness must be considered, he realized; that was what he had previously left out of his calculations.

His fundamental error had been to assume that mind would be transformed through matter. He had focused on the surface, he now realized, on matter and its transformations. But actually matter could be viewed as a symbol or manifestation of the deeper movement, Universal Thought. It should be added that by *thought* Bohm did not mean a particular sort of rationality but feelings, sensations, and the whole content and structure of consciousness. Out of his personal crisis Bohm was discovering a new path—and the vacuum left by the loss of Marxism was being filled.

During the rest of his time in Israel, he read and reread Hegel. Later he would joke that Saral kept asking him, "Why are you reading that book again?" meaning Hegel's *Logic*. "Haven't you finished it yet?"[14]

It would be too simplistic to view Bohm's change of attitude as a straight conversion from materialism to idealism; he was far too subtle a thinker for that. What he was extracting from Hegel was a means of perception, within the mind, of the basic movement of nature. For Hegel, the ground of this movement was thought, which, in its dialectical motion, gives rise to matter. Bohm's own thinking was moving toward a nondualistic position in which mind and matter are aspects, or sides, of a deeper whole. Later in life he would use the analogy of a magnet with its north and south poles. Try to separate north from south—mind from matter—by cutting the magnet in half, and in the process you generate new north and south poles. The resulting halves still possess both poles. No matter how many times the magnet is divided, it is never possible to isolate a north pole.

What was the position of science in his new thinking? No matter where his philosophy took him, Bohm could never really abandon his love of physics. He came alive when he talked about it; he exhaled ideas like pollen from a flower. His absolute faith in the power of rational science may have been shaken by political events, but he still felt that a proper understanding of the nature of matter could help bring human consciousness into order.

In keeping with his passion for Hegel and the dialectic, Bohm now developed an interest in mathematical transformations. He had formerly been distrustful of elegant mathematics and its increasing importance in physics. Now that he had moved from an exclusive consideration of the world of matter to a cosmology that included thought, he had to rethink the role of mathematics in expressing this unity.

Bohm now considered the way canonical transformations work in classical mechanics. A canonical transformation is a mathematical transformation of the equations that describe the behavior of a system of particles, while at the same time preserving a certain form. To use a metaphor, this paragraph can be translated into French, German, Arabic, Japanese, and so on. In each of these languages, the symbols of its alphabet, the look of the words, and even the length of the sentences will be different. Nevertheless the underlying meaning of each translation would be the same. Translating does not involve writing down an arbitrary pattern of words in the new language; rather, it uses the sequence of words that preserves the meaning expressed in the original language. Accurate translations are transformations from one language to another that preserve meaning.

Another example would be a musical canon, in which the theme undergoes a series of transformations, being transposed, inverted, played backward, and so on. While at one level the sound of the theme changes, its essential structure, expressed in the intervals between notes, remains unchanged—it is invariant.

Similarly, the canonical transformations of classical physics produce descriptions that appear, on the surface, to be totally different. Nevertheless, something fundamental in these descriptions remains unchanged or, to use the technical term, is invariant.* Under a canonical transformation it is possible to move from one physical description to another in such a way that the coordinates of one description become enfolded within those of the other.

As a mathematical technique for solving problems in theoretical physics, canonical transformations had been around since the nineteenth century. Bohm was now struck by the fact that a given system can be described in widely different ways. Most physicists simply take these alternative descriptions as mathematical artifacts, but Bohm felt they were pointing to something deeper. In one canonical transformation, for example, it is possible to transform an interacting set of particles into what looks like a system of waves. The surface descriptions are profoundly different, but all the predictions about what can actually be measured—energy, momentum, and other observables—are identical. Bohm realized that it was not necessary to look at quantum theory to find a dual description—waves or particles. Duality was already inherent in classical mechanics.

The idea of a canonical transformation, in which surface descriptions change while the underlying form, the Hamiltonian, remains invariant, evoked for Bohm the discussions he had had with Mario Schönberg in Brazil. He thought back to his cosmology of infinite levels of matter. What is necessity at one level becomes contingent when seen from the perspective of others. Maybe it would be possible to represent this movement by means of a canonical transformation. The way in which, through a canonical transformation, one descrip-

* Canonical transformations preserve the form of what is known as the Hamiltonian of the system. The Hamiltonian contains information about the number and kinds of particles in a system and the forces by which they interact. Under a canonical transformation, the form of this Hamiltonian must remain invariant.

David Bohm. (© by Mark Edwards/Still Pictures)

Frieda Bohm, David's mother.

Samuel Bohm, David's father. The dapper, lady-killer aspect of Samuel Bohm is much in evidence.

The furniture store in Wilkes-Barre (under new management) where David grew up.

Saral Woolfson before her marriage to David Bohm.

A youthful David Bohm.

The Princeton Physics Department. David Bohm is second from the right in the second row.

David Bohm with David Yevick in Loosdrecht, Holland, 1955. (Photo courtesy of Miriam Yevick)

David and Saral Bohm. (© by Mark Edwards Picture Library)

Bohm with the Dalai Lama during the Amsterdam conference "Art Meets Science and Religion in a Changing Economic Environment."

Bohm and Jiddu Krishnamurti in conversation. (Photograph by Mark Edwards © The Krishnamurti Foundation)

The Bohms with relations in Jerusalem.

Bohm in Ojai, California, at his favorite occupation: walking, thinking, and talking, with his wife, Saral, and the philosopher Paavo Pylkkanen.

The Bohms at Michelton with Peter Garret, Jenny Garret, Don Factor, and Anna Factor. During this meeting Bohm developed many of his ideas on dialogue which led to the book *Unfolding Meaning*.

Bohm on his retirement from Birkbeck College in 1983. Standing, from left to right: Allan Appleton, Peter Trent, John Hirsch, Malcolm Coupland, Basil Hiley, and Keith Higgins. Seated: John Jennings, John Hasted, and David Bohm.

Bohm and his hidden variable collaborator, Jean-Paul Vigier.

When speaking Bohm employed delicate movements of his hands, almost as if they were the extension of his thinking process.

Bohm and John E. Fetzer, during one of the physicist's visits to the Fetzer Institute.

David Bohm during the 1980 Amsterdam conference "Art Meets Science and Religion in a Changing Economic Environment." There were times when Bohm appeared totally drained of energy. Yet moments later, if his interest was stimulated, he would talk in a highly animated way.

David Bohm and the photographer, Mark Edwards, while working on their book *Changing Consciousness*. Edwards is an environmental photographer whose work illustrates the extreme degradation of the environment and the changing face of human society. Their book together demonstrated the way in which the origin of such problems can be traced to the operation of thought. (© by Mark Edwards/Still Pictures)

David Bohm's grave in Waltham Abbey.

tion becomes enfolded within another also echoed his treatment of individual and collective motions in his theory of the plasma. Saral became so excited with this cluster of ideas that she made a clay sculpture that expressed the nature of canonical transformations.

Although his attitude toward mathematics was more favorable, Bohm never finally granted it the supreme importance it had assumed in twentieth-century theoretical physics. He was critical of the remark, made by astronomer Sir James Jeans, that "the Great Architect of the Universe now begins to appear as a pure mathematician," and by Heisenberg's argument, referring to quantum theory, that the reality is in the mathematics. Later, in the 1980s, when superstrings came on the scene, Bohm decried the fact that superstring theory was based upon elegant mathematics without any deep, underlying philosophical motivation or sense of physical necessity. Mathematics was an area of thought that had come to order, in Bohm's view, yet it always remained limited and must never be mistaken for the whole. While Bohm had begun to seek a mathematical language that would be appropriate for the ideas he was developing, it was an endeavor that always had to be kept in balance.

In addition to developing his ideas on canonical transformations and a cosmology of infinite levels, Bohm traveled to Paris each summer to pursue his causal interpretation. Bohm may have feared an exhibition of strong feelings, but while working with Jean-Paul Vigier, he was ready to defend the nuances of his interpretation with the utmost vigor. Vigier is an enthusiastic and extremely volatile character, and Saral wondered if the argumentative style Bohm adopted during the late 1950s and 1960s grew out of his interactions with him.

Visits to Paris continued even after Bohm moved from Israel to England, tailing off only at the end of the 1960s. During that decade Bohm lost interest in his hidden variable approach, in favor of questions about structure, process, and order in physics. As far as Bohm was concerned, there were more interesting problems to pursue. Only in the 1970s was his interest in hidden variables revived through the research of his student, Christopher Philippidis.

CHAPTER 11

Bristol: Encounters with Famous Men

ISRAEL AND THE TECHNION had been places of refuge from the United States and Brazil, but soon after his arrival there, Bohm realized that he would have little in the way of scientific stimulation. He got on well with Nathan Rosen, the head of the department and co-author with Einstein and Podolsky of the famous EPR paradox, but the older man's research did not interest him. Instead he looked to Europe. An earlier offer, from Maurice Pryce, of a professorship at Bristol University had expired, but Bohm decided to accept the lesser post of research associate. In the late summer of 1957, he and Saral left Israel. On their way to Bristol, Bohm spent a few weeks at the Bohr Institute in Copenhagen. Niels Bohr invited him to his home, where Bohm attempted to explain his new ideas on cosmology. This, he believed, was a much deeper way of understanding quantum theory. But, as others had discovered before him, it was difficult to hold a dialogue with Bohr. As Bohm tried to explain his ideas, Bohr kept to his own train of thought or made statements that did not seem particularly clear.

This behavior should have come as no surprise to Bohm, for the physics community abounded with anecdotes about Bohr, such as his belief that while the opposite of truth is an error, the opposite of a great truth is another great truth. His collaborator Leon Rosenfeld claimed that this maxim even influenced the way Bohr wrote his scientific papers. Bohr would agonize about the best way to express a particular idea in a given sentence; then, when he had written it down, he would attempt to express its exact opposite in the sentence that followed.[1] One story, which may be apocryphal but sums up Bohr's character, tells of the time Bohr and his wife argued for many hours about which baby carriage to buy. In the end, Bohr's wife caved in and agreed with her husband's choice. This only made Niels Bohr unhappier, for, although she agreed with him, he felt it was not for the right reasons.[2]

Bohm, for his part, noticed Bohr's annoying habit of arresting all conversation while he filled his pipe. On occasion he would drop his box of matches and then very carefully pick up each match before lighting his pipe. Bohm felt that Bohr used the tactic to deflect the conversation from his opponent's track and back toward his own.*

If Bohr tended to keep to his own train of thought, David Bohm could do the same, single-mindedly pressing his own point of view. One imagines that the two physicists made little headway in reaching a common accord. From Bohm's point of view, Bohr was insistently and passionately convinced of his own position to the extent that he showed little interest in other people's ideas. But on a more positive note, he discovered that, unlike so many contemporary physicists, Bohr had great strength when it came to philosophy. In the end, however, it was difficult for Bohm to convince the Danish physicist of the value of his ideas, which Bohr found "beautiful" but "not on the right track."[3]

If Bohm and Bohr had no direct encounter of minds, at least Bohm developed a great respect for the elder statesman of quantum theory. He discovered Bohr's thinking to be particularly subtle—another factor that made it difficult to argue with him. And if Bohr was inflexible in his adherence to the Copenhagen interpretation, then at least he was uplifting to talk with. With the notable exception of Einstein, most of the other physicists Bohm had talked to were interested only in details. Bohr, by contrast, was concerned with deeper

* Another anecdote, told to me by my thesis supervisor, Tom Grimley, tells of how, after a seminar, Bohr was asked a particularly penetrating question. The great man attempted to light his pipe, dropped his matches, and spent a painfully long time picking up each one and individually returning it to the box. Finally he lit his pipe, looked around the room, and asked, "Now, are there any questions?"

issues. He had an exceptional ability to sort out physical problems and difficult ideas, and he was willing to be helpful and encouraging to others. Bohm went so far as to describe Bohr as "inspiring."*[4]

Despite their differences, Bohm realized that Bohr was one of the very few physicists who was trying to understand the meaning of physics at a truly fundamental level. When he formulated the Copenhagen interpretation, Bohr had been influenced by American psychologist William James's notion of the "stream of consciousness." The act of observing a thought in detail changes the nature of that thought, and as with observer and observed in quantum theory, so the thought and the thinker can never be separate—they form a totality. Bohm felt a particular sympathy with this insight, since he had always believed that his own mental processes reflected this totality; moreover, he believed that the things he was thinking about were actually taking place within his own body and mind. For Bohm, as for the medieval philosophers who adhered to the maxim "as above so below," the nature of reality could be directly apprehended within one's being—a notion that conventional physics hardly touched on at all.[5]

While he was in Copenhagen, Bohm had an insight into the nature of infinity, an issue he had been thinking about for some years. The vision came to him in the form of a large number of highly silvered spherical mirrors that reflected each other. The universe was composed of this infinity of reflections, and of reflections of reflections. Every atom was reflecting in this way, and the infinity of these reflections was reflected in each thing; each was an infinite reflection of the whole. This image possesses almost mystical connotations in its vividness, and at the metaphorical level at least, it contains the seeds of the concept that Bohm was later to call the implicate order.

The Bohms' next stop was Bristol University, where David Bohm

* Bohm's attitude toward Bohr reminds me of that found in the partners of a long-term relationship. While Bohm was critical of what he felt were the limitations of Bohr's interpretation of the quantum theory, he would jump to the great man's defense whenever anyone else criticized him. I think this paradoxical behavior stemmed from the fact that Bohm had spent a great deal of time trying to understand Bohr and, from this position, felt that he had the right to criticize the Danish physicist. However, the vast majority of physicists, while paying lip service to Bohr, generally had only a partial or distorted conception of the Copenhagen interpretation. Bohm was therefore at pains to set them straight.

was soon complaining about the difficulty of getting any work done. Although he had grounds for this grievance, by now one can anticipate Bohm's protests at each new location, as if they were an essential part of his nature. The head of the physics department, Maurice Pryce, had been a member of the British contingent working at the Radiation Laboratory in Berkeley during the war. At that time Bohm got along well with him, feeling him open and democratic. But in Bristol he found the situation far from ideal. Too much emphasis was placed upon status, he felt, even to the seats people occupied during seminars. A departmental coffee or tea break is generally a time for open talk, but at Bristol Bohm found the conversation strained, and he sensed that people were constantly watching each other. He also resented the fact that he was being paid at the very bottom of the scale attached to his research associateship. After he complained, Pryce told Saral that while Bohm may have been a big fish in the small pond of Israel, in Bristol he was just a little fish in a big pond.[6]

On the whole, however, the situation was not all that bad, for the Bohms hosted a fortnightly group that discussed the philosophy of science. David enjoyed discussions with the physicist Herbert Fröchlich and his wife, Fancon, who was an artist.* He also had the company of his two Israeli students, Yakir Aharonov and Gideon Carmi. His energies were also revived by interactions with the philosophers Stephan Körner and Paul Feyerabend. Feyerabend, who had previously worked with Karl Popper, had an interest in physics and philosophy that made life more tolerable for Bohm. (Feyerabend had been present in Copenhagen when Niels Bohr first had word of Bohm's hidden variable papers.)

Feyerabend was struck by the intensity of Bohm's discussions. On one occasion Bohm called at Feyerabend's home, walked into the

* Many years later I asked the Fröchlichs if they had noted a change in Bohm's thinking as the result of his encounter with Krishnamurti. Many people who knew Bohm after he met Krishnamurti believed that the Indian teacher had exerted an enormous influence on the physicist. But Fancon Fröchlich did not think that there had been any essential change in Bohm's thinking. She noted that at Bristol Bohm had been extremely interested in the process philosophy of Alfred North Whitehead and that the general language Bohm used then, and in the late 1980s, was essentially the same.

living room, and took off his raincoat, all the while enthusiastically discussing philosophy, only to find that Feyerabend was not home![7]

With the Bohms in England, Saral's relatives had plenty of opportunity to meet her new husband. As it turned out, they considered him far from promising. Sometimes they visited Saral's Uncle Alfred Randolph, deputy director of a department in the Ministry of Health and Pensions. Uncle Alfred discovered that Bohm did not behave in a normal way, sitting with the family at night and exchanging small talk. Instead he spent all his time in the bedroom, to the point where Uncle Alfred was moved to inquire, "What's he doing up there?" Saral replied that he was simply sitting. "But doesn't he do any work?" Uncle Alfred asked. "That is his work," she explained.[8]

On one occasion it looked as if Bohm had turned over a new leaf. He and Saral were taking a break at the seaside resort of Weston-super-Mare. While walking along the beach, Bohm had an insight into a way of making unlimited energy. Excited, he asked Saral to telephone her uncle, who was familiar with official procedures and could help him secure a patent and develop his discovery.

With plans afoot to produce unlimited energy, it looked as if the new family member were at last showing common sense. Even Uncle Alfred approved, saying, "Now at least the man's doing something." The couple planned how they would spend the vast fortune they would make. Bohm was going to build an institute for scientists, artists, musicians, and philosophers, where the best brains in the world could work together creatively and engage in genuine exchanges without ever having to worry about money or competition. A few days later he discovered his idea about free energy was wrong and he announced, "It's not going to work." At least, Saral thought, it had been great fun spending their money in advance.[9]

If unlimited funding never materialized, Bohm's vision of a group of scientists working freely together never faded. In the 1980s he still talked of organizing a meeting where scientists would come simply to talk about "doing science" and not feel they had to present or justify their ideas.

Bohm's arrival in England coincided with the publication of *Causality and Chance in Modern Physics*.[10] Despite good reviews, some of which argued that it was an important work, it did not sell particularly well and eventually went out of print. *Causality and Chance* was the distillation of Bohm's thinking during his time in Brazil. Now he could see how, in its approach, it had fallen between the two stools of physics and philosophy. Wishing to get to the heart of the problem as quickly as possible, Bohm had not included or even acknowledged the historical discussion of causality, from Aristotle to Kant and Hume. This omission may have led philosophers to dismiss his approach as yet another example of a physicist's naive realism. For their part, physicists may have been disappointed to find nothing of immediate, practical use. In any event, Bohm was disappointed by the reception and concluded that physicists were no longer interested in the wider philosophical implications of their work.*

While he was at Bristol, Bohm also completed an undergraduate textbook, *Modern Physics,* which never saw publication. The three publisher's referees enjoyed parts of the book, but their general criticism was that it moved between an elementary approach and high-powered mathematics. It was not at all clear who would be the audience for such a book. As a result, the manuscript was filed away and rediscovered only after Bohm's death.

While he was working with Gross and Pines at Princeton, Bohm had developed a new description of the plasma, in which the coordinates for individual and collective are intertwined—free individual motion was enfolded within the collective and vice versa. The plasma, through its complexity, developed a new order, a cooperative oscillation involving an astronomical number of electrons.

Ever since his schoolboy theorizing about the cosmos, Bohm had been interested in how order can emerge out of an underlying ground, be it an ether or a collection of particles. The plasma was one such example, in which collective oscillations emerged out of an astronomical number of individual motions. Most physicists believe that space and time are prior—givens in which matter moves and has its being. Likewise, the laws of physics exist a priori and exert their influence on matter. But what if space and other structures emerge in a similar sort of way, out of a flux or complex activity that exists before space and time? It could be that coupling together underlying pro-

* The book was later reprinted as Bohm's ideas reached a new and wider audience.

cesses gives rise to that averaging-out we call space. This underlying ground may have a highly complex order, even the appearance of chaos. Therefore what we know as the present laws of nature may emerge as some sort of averaging procedure.

Bohm remained interested in this general question of the emergence of structure and order out of an underlying ground. With Gideon Carmi he explored such a model, beginning with a collection of particles and the conventional laws of classical physics. Bohm and Carmi showed how, in an analogous fashion to the plasma, the description of a system of very many particles can be transformed into a new dual description consisting of quasi-independent motion within a large-scale, global structure. Unfortunately these papers never received the recognition they deserved, despite the fact that they contain valuable insights and anticipate approaches that were later to become fashionable in chaos theory (or more accurately, nonlinear dynamics.)

Bohm's work with Aharonov was quite different. The student had become interested in what physicists call the vector potential, and Bohm immediately recognized the novel and unexpected features inherent in Aharonov's conjecture. Since the work they produced has been described as of Nobel Prize stature, it is worth spending a few moments to explain the nature of the discovery.

The phenomena associated with magnetism and electricity have been known for centuries. In the early nineteenth century, physicists noted how changes in a magnetic field—such as a magnet moving past a coil of wire—can produce electrical effects; likewise, a fluctuating electrical current was known to exert an effect upon a magnetic compass needle. Clearly the phenomena of electricity and magnetism were in some way related. The important theoretical insight came later in the nineteenth century, when Maxwell introduced the concept of fields into physics—electrical fields and magnetic fields—and wrote down the equations that governed them. In Maxwell's equations, as they became known, the electrical and magnetic fields are interrelated through what is called the vector potential. As every physics student was taught, this vector potential has no physical reality in itself; it is not something that can be measured or detected. Rather, it is an abstract concept, a mathematical tool, used to relate one field to another.

Aharonov questioned this basic assumption and, with Bohm's encouragement and clarity of mind, attempted to show how the presence of this potential could be made manifest. In the famous double slit experiment, a single electron appears to pass through both slits before encountering a photographic plate. The pattern produced by averaging very many such events shows interference fringes. From this it appears that a single electron is capable of interfering with itself. (Of course the language of "paths" and electrons interfering with themselves is strongly rejected by the Copenhagen interpretation. It is used here as a shorthand way of presenting the Aharonov's conjecture.)

Suppose that an electrical field exists behind these two slits, but is shielded in such an way that, after traveling through the slits, the electron never encounters this field. Classical physics dictates that the presence or absence of a shielded field should have absolutely no effect on the subsequent interference pattern.

Aharonov and Bohm pointed out that even if the electrical field is zero—that is, totally shielded from the electrons—the vector potential is not. Of course, classical physics denies the reality of the vector potential, claiming that it is no more than a mathematical device used to aid calculations. Aharonov and Bohm proved otherwise and argued that the amount by which the interference pattern shifts is related to the amount of flux trapped in the electrical field—even though the electron never passes through that field. Aharonov and Bohm had made explicit the essential global nature of quantum theory. Just as Bohr had argued, the whole of the experimental arrangement must be taken into account when defining a quantum state. Now it could be shown that even parts of the apparatus that, in a classical way, are totally shielded from the electron nonetheless have significant effects at the quantum level. (The more technically inclined reader should know that Aharonov and Bohm's work is the first example in physics of a gauge field.)

The possibility of observing the effects of the vector potential and demonstrating the global nature of quantum phenomena came to the attention of experimental physicist Robert G. Chambers, who occupied the office next door to Bohm, or rather the same room with a lightweight partition dividing the two halves. Over the past months

Chambers had been forced to put up with the constant background noise of Bohm jingling coins as he paced to and fro. From time to time this sound was interrupted by Bohm's highly animated discussions with his two students, Aharonov and Carmi. It was said that they used to tape-record their brainstorming sessions in case anything interesting came up that they might forget.*

Chambers worked on Saturday mornings, and just as he was about to leave for lunch, Bohm had the irritating habit of arriving to tell him all about his latest ideas. Most Saturdays Chambers's mind was too much on his stomach to benefit from the lectures, but once the Aharonov-Bohm effect had been discovered, it was clear that someone had to test out the idea. There was an old electron microscope in the department, and Chambers took up the challenge of modifying the machine so that the idea could be tested.

Chambers was able to demonstrate what Aharonov and Bohm had theoretically predicted—a totally new effect in physics. Unfortunately, the apparatus Chambers used was not the best possible for that particular experiment, and it could be objected that some of the electrical field was leaking out and affecting the electron. Since the Aharonov-Bohm effect was so important and so counterintuitive, it was important to rule out all objections. Over the years other experiments were designed, even to the extent of shielding the electrical field by means of a superconductor. Ultimately the Aharonov-Bohm effect came to be accepted. Yet to some physicists it remains so surprising even today that papers sometimes still appear attempting to demonstrate that the effect does not, in fact, exist.[11]

There is an interesting corollary to the AB effect, as it was then known. After their first paper had been published, Bohm learned that the effect had already been postulated by a maverick physicist called Rory E. Siday. In the late 1940s Siday had been working at Birkbeck College with Werner Ehrenberg on electron lenses, when he noticed a bizarre effect. Ehrenberg persuaded Siday to write it up as a

* Aharonov had an exceptionally intelligent mind. When it came to his Ph.D. orals, Maurice Pryce and Rudolph Peierls acted as examiners. It was said that Aharonov had thought up a few difficult but intriguing questions that he dropped casually into the conversation. This set his two examiners arguing with each other while Aharonov sat back to listen.

scientific paper, which he did in a somewhat old-fashioned way, without really presenting a proper quantum mechanical account of the phenomenon.[12]

Another physicist at Birkbeck, David Butt, who had been a student of the great Max Born, suggested that Siday talk to Born himself about the idea. The meeting took place in a very large room, Siday sitting at one end of a big table and Born at the other. The atmosphere rapidly became heated and angry, with Siday saying that there was a genuine puzzle to be resolved and Born saying that no problem existed at all.[13]

When the paper, to which Ehrenberg added his name, was published, it fell flat and was not taken up by the physics community. This explains why, in the era before computer searches of scientific papers, Aharonov and Bohm had not come across it. They did, however, add a reference to Siday and Ehrenberg in their second paper, also written in Bristol, and when Bohm and Basil Hiley later wrote on the same subject, they named it the ESAB effect, rather than the AB (Aharonov-Bohm) effect. Certainly, as Hiley has since pointed out, it was out of the question that Ehrenberg had ever felt unjustly treated by Aharonov and Bohm; if he had, he would hardly have been party to inviting Bohm to Birkbeck or so actively supported him on the appointments committee. Indeed, Ehrenberg was happy that this earlier work had been taken up again.

The work of Aharonov and Bohm was considered by many physicists to be of Nobel Prize quality, and over the years rumors surfaced that they were short-listed for the prize. But no award was ever made—possibly, physicists speculated among themselves, because of the ambiguity over who exactly had discovered the effect.

In Bristol Bohm continued his work in physics, yet at the same time the feeling grew within him that there was more to the world than physics could explain, or rather, that there had to be something qualitatively different. If human society were to be transformed, he sensed, the processes involved had to reach even deeper than Marxism had allowed. His cosmology of qualitatively different levels had to include a place for consciousness.

The result was that he cast his net even wider, visiting the local library and bringing home books on the Christian mystics, Indian philosophers, yoga and Buddhism. In particular, he came across the work of P. D. Ouspensky and G. I. Gurdjieff. Gurdjieff had taught that people are driven by forces and reactions that are largely unconscious. When everyone behaves like a somnambulist, where is human freedom? Freedom, Gurdjieff wrote, is an illusion. Our choices are determined by largely underlying irrational impulses. The burning question Gurdjieff posed was how to awaken people from their sleep.

Bohm was struck by the story of how P. D. Ouspensky, an important disciple and interpreter of Gurdjieff, had watched people on the London underground and compared them to sleepwalkers. In his own life Bohm was painfully aware of the irrational and impulsive ways in which people behave. He compared the human race to people asleep on runaway horses. The problem was not so much to control the horse as to wake up the rider, then to have him or her stay awake long enough to realize the danger.

If Gurdjieff had stated the basic problem facing human beings, what was the next step? How was the sleeper to awaken and achieve true freedom? To Bohm, the solutions Gurdjieff had proposed were mere psychological tricks. One approach had been to place his disciples in absurd or stressful situations. He would, for example, serve an elaborate meal and press food and drink on disciples long after they were satiated. Because the disciples held Gurdjieff in such respect, they would not refuse what was offered, even beyond the point of personal discomfort. In this way Gurdjieff hoped to awaken in them the realization that they had no will of their own. If they could be awakened for only a moment, they might be able to look more carefully at their own reactions. Operating through confrontation, paradox, and an extraordinary form of dancing, the Master would alert his disciples to their powers and potentials.

Bohm did not think much of this project. It was unlikely that he could be persuaded to gyrate wildly to music and fling himself from a stage into an orchestra pit. If Bohm were ever to awaken on the runaway horse, he would have to do it in a more sedate and contemplative manner. Still, Gurdjieff had raised the problem that Bohm was now to address with all his energies.

When his former colleagues in the United States learned of this change of interest, it caused them considerable distress. In the years that followed, some lamented that Bohm had gone "off the rails," that a great mind had been sidetracked, and the work of an exceptional physicist was being lost to science. Yet in the context of Bohm's childhood dreams and visions of light and of vast energies, his fascination with the "ultimate," and his stories of highly evolved consciousnesses living on distant planets, his new path was consistent with everything that had gone before.

The catalyst for what was to be a major transformation of his life and work was discovered by Saral Bohm during one of their visits to the public library. There she came across a book that contained the phrase "the observer is the observed." This sounded to her exactly like the sort of thing David was always talking about in the context of quantum theory.

Saral showed him *The First and Last Freedom,* written by the Indian teacher Jiddu Krishnamurti. Bohm read the book as fast as he could, then borrowed more books by the same author. Here was a thinker who had seen deeply and authentically into the essence of the human problem. Gurdjieff had warned of the trap of unconscious conditioning; Krishnamurti was pointing to a way out. He wrote of the transformation of human consciousness through the operation of the "intelligence," or the "unconditioned."* Having long dreamed of transcendence, now a doorway to what lies beyond thought was being opened for Bohm. Bohm wrote to the American publisher to discover if Krishnamurti was still alive and if additional books were available. He learned that in June 1961 the Indian teacher would be making his first visit to London in several years. He would hold a series of talks, but no individual meetings were possible.

When the time arrived, the Bohms traveled to London and stayed at a small hotel. At the first talk Bohm would have seen a fine-boned Indian, impeccably dressed in the best that Savile Row had to offer. His features were handsome and delicate, a face that lit up in anima-

* In his early writings Krishnamurti used words in specialized and often poetic ways. Krishnamurti's term *intelligence* should not be confused with the purely rational functioning of thought. It implies something that lies beyond discursive thought and that acts when the mind is silent and has died to thought. He taught that this intelligence is able to bring about a mutation, or physical transformation, of the brain itself.

tion as he spoke, hands gracefully employed to emphasize his words, eyes at one moment soft and compassionate and, at the next, burning with passion. His talk would begin in a tentative fashion, as if he were probing into some new area for the very first time. Possibly he would invite his audience to suggest a topic and then begin tentatively, like a connoisseur handling an exceptional piece of porcelain, gently turning it in his hands, commenting on its beauty, pointing out singular features, inviting his audience to participate in his enjoyment rather than offering a dogmatic opinion.

Krishnamurti did not teach in the traditional sense of imparting knowledge, referring to dogmas and ancient traditions, seeking adherents and disciples, suggesting disciplines and practices, and emphasizing a conclusion. Rather, he asked his listeners to accompany him on a journey. His topics were all-encompassing—suffering, the nature of freedom, the ending of time, "dying to the moment." He proceeded via a series of questions, asking in effect "What is this thing? Could it be this? Could it be that?" His technique, if that is the correct word, was the *via negativa,* showing the futility of the approaches and answers that other teachers drew upon.

As he spoke, he invested these burning questions with such passion and intensity that his listeners were drawn into the quest, seeking in their own minds a way to answer the questions and each time realizing the inadequacy, indeed the total futility of their answers. In this way each person was brought to the edge, face to face with the abyss. Responding to Krishnamurti's questions, they traveled into a new terrain and reached that point where the next step would take them into the void. Krishnamurti spoke of the transformation of consciousness, and his audience realized that, no matter how brilliant they might be in physics, law, philosophy, or painting, they were unprepared to make that leap.

Krishnamurti had brought his audience to the perception that whatever they planned, believed in, hoped, or thought about could in no way meet the case. Something totally different was called for. There was great psychological resistance to this perception, each mind seeking a way of escape. But was it possible to remain with this tension, Krishnamurti asked, to suspend the desperate need for action, engage with such energy that something different came into existence? Again and again those who attended Krishnamurti's lectures found themselves suspended at this point. It should be added that in their personal lives, many of them fell into a sort of paralysis, one in which any form of practical action seemed inadequate.

Being taken to the edge must have been a particularly challenging experience for Bohm. In his physics he had always attempted to go to the limit, constantly seeking the question that lies beyond the question. He had dreamed of transcendence, of what lies beyond the everyday world of shadows. Now he was listening to someone who claimed to voyage and even live in the world beyond that edge. It was of the greatest importance to Bohm that he should talk to him face to face, and he wrote to the London organization. The result was a meeting at the Wimbledon house where Krishnamurti was staying.

Bohm knew little of the personal life of the man he was about to meet.[14] Only later did he learn that Jiddu Krishnamurti's origins as a teacher (on the earthly plane, at least) went back to that remarkable woman, Annie Besant, who in her earlier years had been a Fabian, a pioneer of women's rights, and a staunch advocate of birth control. After coming upon Madame Blavatsky's "Secret Doctrine," Besant experienced a religious conversion.

Helena Petrovna Blavatsky had founded the Theosophical Society to teach her syncretic mixture of oriental philosophy and religion. In 1907 Besant became the society's president, adding her own brand of esoteric Christianity to the mixture. Theosophists believed, among other things, in figures who guided the spiritual evolution of the human race, and they spoke of the coming of the World Teacher. It was said that this being had suffered many reincarnations in order to reach a state of perfection and that even now a physical body was being prepared for him.

Annie Besant became friendly with Charles Webster Leadbeater, who had earlier been expelled from the Theosophical Society. Convinced of his clairvoyant powers, she had him reinstated, and the two traveled to India, where Leadbeater spotted the son of Jiddu Naraniah, a minor functionary in the Theosophical Society. Leadbeater watched the boy bathing in the river with his companions and was struck by his aura of total selflessness. As he probed into the boy's earlier reincarnations, he became convinced that young Jiddu

Krishnamurti was the vehicle for the Bodhisattva Maitreya. Leadbeater and Besant prepared the boy for this manifestation, and while his education was generally conducted along traditional English lines, he was also accepted for instruction on the astral plain by Master Koot Hoomi—all of which caused some controversy within Theosophical circles and precipitated the apostasy of Rudolf Steiner, who went on to form his own Anthroposophical Society.

Krishnamurti's first initiation on the astral plane took place in January 1910, and in the years that followed, he lived in England, advancing through a series of spiritual levels until he reached the point where Lord Maitreya would speak through him.

During this period Leadbeater also spotted D. Rajagopalalacharaya, whom he predicted would become the Buddha of Mercury—admittedly a more modest-sized planet than Earth but one that nonetheless demands a liberal application of sunblock. Rajagopal, who was later to play a significant role in Krishnamurti's life, was sent to England, where he and Krishnamurti met, initially with some reserve. While Rajagopal excelled at his schooling, Krishnamurti failed his examinations and appeared in many ways to be a vacant young man.

But on the astral plane Krishnamurti was moving through spiritual levels much faster than Leadbeater had believed possible. Soon he began to speak in the first person when referring to Lord Maitreya. He claimed not to be the vehicle but to have merged his consciousness with that of "the beloved"—an even more highly evolved spiritual entity than Lord Maitreya.

On several occasions each year, Krishnamurti addressed his followers at large-scale meetings, and while speaking before a large audience at Ommen, Holland, on August 3, 1929, he performed a particularly dramatic act: He formally dissolved the Order of the Star, which had been created around him, with the words "Truth is a pathless land." Human freedom could not be gained by belonging to an organization, practicing a religion, or following a guru, he said. From now on, Krishnamurti would travel the world, giving talks and answering questions. Rajagopal would supervise his various travel and publishing projects.

In the 1920s and 1930s, Krishnamurti had a distinguished following, including such figures as the writer Aldous Huxley. Years later, on hearing Krishnamurti speak at Gstaad, Switzerland, in 1961, Huxley wrote, "It was like listening to a discourse of the Buddha—such power, such intrinsic authority."

In his public talks Krishnamurti spoke of "dying to the moment," of human freedom, and the end of suffering. He hinted at the transformation of consciousness that can take place in a mind that is totally silent. In his own case it appeared that the irrational conditionings that entrap the rest of humanity had never "stuck" with him.

In 1961 he began to record his states of consciousness and write of a mysterious "it," or benediction, that came to him each day in a palpable way. "The process," he wrote, was accompanied by such intense pain in his head and spine that at times he would faint. From the 1920s through the 1940s, he had spoken of Krishna and others visiting him, as if his body and brain were vehicles to be used by other powers.

Into this heady and esoteric world, Bohm entered passionately and wholeheartedly. In Krishnamurti's books he found a clear analysis of the nature of consciousness and the mechanism whereby the thinker separates him- or herself from the thought and the action of thinking, by positing him- or herself as a separate, independent entity. In this act of separation, and in the subsequent reification of the thinker and the thought, lie the origins of human problems. Krishnamurti's observations that "the thinker is the thought" and "the observer is the observed" struck Bohm as resembling his own—and Niels Bohr's—meditations on the role of the observer in quantum theory. Bohm had personally experienced the way in which the observation of a particular thought changes the movement of thought itself. His study of Hegel had led him to similar conclusions about the movement of thought. The physicist was well prepared for his engagement with Jiddu Krishnamurti.

The meeting was important for both men. Before World War II the majority of Krishnamurti's audiences had been connected with the Theosophical Society and were familiar with the language and context in which he spoke. In the 1950s the new audiences were less able to follow his teachings. The world had changed, and Krishnamurti knew that he had to adapt his approach. His discussions with David

Bohm therefore came as a particular challenge. Bohm was one of the very few people in the West who could hold a sustained dialogue with Krishnamurti and, in the process, help him to refine his approach to communication, making it less poetic but more precise in the use of certain words.

All this lay in the future, for at their first encounter the two men sat in absolute silence. While it may be unusual to remain quiet in someone's company for so long, Bohm felt an absence of tension. In the end it was Saral who broke the silence, suggesting that Bohm should explain his work to Krishnamurti. Although the Indian teacher would not have understood the technicalities of Bohm's research, he listened attentively and appeared to grasp the spirit of what Bohm was saying.

As Bohm spoke, he had a feeling of intense communication with no holding back. It was the same feeling of energy, openness, and clarity that he had sometimes experienced when speaking with scientific colleagues who were vividly interested in his ideas. At one point Bohm used the word *totality,* whereupon Krishnamurti jumped from his chair and embraced the physicist, saying, "Yes, that's it. Totality."[15]

The meeting was everything Bohm dreamed of. Krishnamurti was totally open and able to go into things with great passion. Bohm compared him to Einstein in his ability to explore deeply in a spirit of impersonal friendship.

Not only had Bohm become disillusioned with Marxism, he had failed to find a community of physicists who were willing to think passionately about the implications of their own subject. The search for transcendence remained one of the most important quests of his life, and he was totally open to the encounter. Following their first meetings Bohm devoted more and more of his time to investigating Krishnamurti's teachings. Their interaction gave him a perspective from which to question the value of his own research, and at times he even contemplated abandoning physics in favor of a total commitment to the Indian teacher.

Bohm's apparent new direction again caused considerable distress among his former colleagues in the United States. For his part, he was finding it increasingly difficult to relate to what other physicists were doing. A particular example came during an international conference on plasmas in Utrecht, in the Netherlands, where he anticipated learning what had been happening to the plasma theory he had developed at Princeton.

According to a probably apocryphal story, Einstein once complained that after the mathematicians had got hold of his theory of relativity, he could no longer understand what he had done. The situation was not dissimilar in the case of plasma theory. At Utrecht Bohm discovered that while the younger generation of physicists were using elaborate mathematical approaches, the underlying physics was shallow. As far as he could see, they only wanted to write down formulae on the blackboard. Many of them spent their time manipulating what are known as Feynman diagrams, as if the diagrams themselves represented physical understanding. The equations seemed to have become a means of communication, with no physical ideas behind them. In addition, the whole enterprise had become highly competitive, with everyone watching everyone else and thinking of their own personal advancement.

Bohm's talk, on the collective and the individual in physics, puzzled the audience. Here was a physicist who had done great things in the past, they must have calculated silently; maybe it would be politic to be associated with him. On the other hand, what he was saying did not seem important, so maybe his influence no longer counted.

Bohm's own work on plasmas had in fact been transformed, he learned, and his former collaborator, David Pines, was now being hailed as an important theoretical physicist. Bohm believed that the physics community thought the most important work on plasmas had been done by Pines alone, rather than emerging out of their intimate collaboration. Bohm was nagged by the suspicion that this misconception had its roots in a change in the order of names on one of the papers, and because of Pines's continued interest in the topic.

In Brazil Bohm had been cut off from the scientific mainstream, his work on hidden variables either rejected or ignored. Now he felt that he was not even being given full recognition for his earlier work. So troubled was Bohm that he brought the subject up again, nearly thirty years later, while tape-recording his reminiscences. When he was engaged in research, Bohm could be generous and selfless, but paradoxically, he also had a need for recognition and approval that dated back to his childhood. Some years after this incident, he told his friend David Shainberg, a psychiatrist, about his inner need for approval and recognition yet, with his colleague, Basil Hiley, he appeared totally indifferent to the opinions of others. Sometimes, Hiley thought, Bohm's confidence in his own approach neared arrogance, and he appeared to make it a virtue to disagree with the scientific mainstream.

The issue centers less on objective fact than on personal interpretation and feelings. Bohm believed that the bulk of the plasma research had been completed when he left the United States. All that remained was for scientific articles to be written. The first paper in the series appeared in the *Physical Review* as written by "Bohm and Pines." But with Bohm in Brazil, Pines was applying for university positions, and as Bohm told the story, he wrote asking if it would be all right for him to put his own name first on one of the papers. It would give him more credit when he applied for a job, he said.

The ordering of names on a scientific paper can lead to heated debate between even the best of friends. Some researchers adopt the neutral convention of listing names in alphabetical order—which is clearly advantageous to those whose names appear near the start of the alphabet. Others hold that the name of the key researcher should appear first. When a series of papers is published and the name order changes during the sequence, it is considered an indication of who took scientific precedence in that particular paper.

What happened next is not too clear. In taping his reminiscences to Maurice Wilkins, Bohm said that he had been so worried about conditions in Brazil that he could not remember replying to Pines's request. Pines, for his part, believed that he had Bohm's agreement, and so "A Collective Description of Electron Interactions: II. Collective vs. Individual Particle Aspects of the Interactions" bore the names "Pines and Bohm," which indicated to most physicists that the majority of this work was performed by Pines.[16] The third paper in the series had names in the order "Bohm and Pines," while the fourth bears the name "David Pines" alone.

David Pines, looking back on this period, agreed that most of the research had been completed before Bohm left for Brazil, but afterward he and Bohm kept up "an active and lengthy correspondence" about the third paper in the series. As to the second paper, in which Pines's name appeared first, Pines believed that they had discussed the order of names *before* Bohm left for Brazil. In Pines's opinion, since more than half of the research had been his own, and since the entire series of papers was based on his Ph.D. dissertation, it was not unreasonable that he should gain recognition for his work.[17]

Bohm and Pines each saw the extent of his own participation in a different light, and as far as Bohm was concerned, the Utrecht conference was another nail in the coffin of physics. Not only was he not given full credit for his plasma work, but most physicists appeared uninterested in the deeper philosophical questions of their subject. To make matters worse, they even ignored the underlying physics they were studying, preferring the surface brilliance of mathematical techniques. In most universities students were now taught quantum theory simply by copying down Schrödinger's equation from the blackboard and then learning how to perform calculations using it. Increasingly Bohm found himself out of sympathy with the mainstream. His own interests were now quite different, and others had even come to regard him as a maverick.

Bohm liked to relate an anecdote about a seminar he once gave in Tel Aviv. At this seminar his audience was bewildered to discover they had missed the meat of his talk. In most talks a physicist introduces the topic in a general way, spending five or at most ten minutes on the overall approach, maybe even mentioning something of the philosophy of the subject. Then he or she gets down to the serious business of writing equations on the blackboard. During the introduction the listeners are preoccupied with finishing their coffee and only come to attending when the real talk begins—that is, when the equations appear.

In Tel Aviv Bohm talked for the customary ten minutes, then twenty minutes, and on into forty minutes without ever writing anything on the blackboard. The audience kept waiting for the real lecture to begin, and only in the last minutes of Bohm's talk did they realize that he was not going to produce any equations and that what they had taken as mere introduction had been his entire lecture!

Again the paradoxes within Bohm emerge. When he was talking

to Krishnamurti, he could question whether science would ever have a significant effect on human consciousness and wonder if he should abandon the entire enterprise. But as soon as he began to talk physics with a colleague, his passion would mount until he became totally absorbed.

Yet now, on so many occasions, Bohm felt he could do nothing right in the eyes of his fellow physicists. The Utrecht meeting had disillusioned him; most physicists did not want to listen to philosophy. Therefore, when it came time to write up his work with Gideon Carmi that had showed genuine insights, he decided to write the paper using a strict mathematical form so that it would appeal to physicists. But not even this strategy worked. It was not just a matter of using mathematics in place of words—he should have employed, it seemed, the particular mathematical language that was in fashion at the time. By an ironic contrast David Pines, on recasting the work on plasmas in the language of Feynman diagrams, had given it an immediate appeal to the younger physicists.

At Bristol Bohm still thought hard about reconciling Einstein's theory of relativity with quantum theory, an issue that remained intractable after several decades of work by the best minds in physics. Most physicists who wrestled with the problem thought of developing alternative theories and new ideas; Bohm realized, by contrast, that something truly radical was needed.

To give one example: A fundamental notion in relativity is the signal. Light signals are used to measure the structure of space-time, and a light signal between two points defines the basic invariant of the theory.* For Einstein's theory to have meaning, a signal has to be a clear and well-defined concept. But in quantum theory the notion of a signal is radically different.

In order to define a signal, one must have two independent, well-defined objects—a sender and a receiver—located at different space-time points, then assume a well-defined interaction, or transmission, between them. But in quantum theory, observer and observed are irreducibly linked. The quantum system is a unified whole that cannot

* This space-time interval is invariant under the Lorentz transformations of special relativity.

be analyzed into a separate sender and receiver. In other words, not only do signals play no fundamental role in quantum theory, they even appear to be at odds with the deeper meaning of the theory. Signals are but one concept that is fundamental to one theory but incommensurable with the other.

For Bohm, the task of reconciling the two was not a question of making modifications to one theory in order to bring it in line with the other. Neither did the situation call for "new mathematics" or clever new ideas. Rather, physics had to pause and delve deeply into the underlying order of these two cornerstones of twentieth-century physics. The significant issue was to discover an entirely new order to physics.

Bohm picked, as a particular example of the need for a radical change of order, the continued use of Cartesian coordinates. In this mathematical description, every point in space is defined by three numbers—its x, y, and z coordinates. The Cartesian order, Bohm pointed out, lies at the heart of Newtonian physics. But Newtonian physics had been overthrown, first by Einstein's relativity and then by quantum theory. Nevertheless both theories retained the same basic Cartesian order—a continuous background of space defined in terms of coordinates on a grid.

Bohm argued that the Cartesian order was incompatible with the new insights of quantum theory. The Cartesian grid is abstract and arbitrary; it has no deep, ontological connection with the quantum world. Moreover, he said, it assumes the existence of infinitesimal points within a continuous, infinitely divisible space. This assumption is incompatible with the basic insights of quantum theory, in which the notion of space breaks down long before the domain of the dimensionless point is reached. A revolution had occurred in physics, but at a deeper level the same order prevailed. The new wine of quantum theory had merely been put in the old bottles of Cartesian order.

Examples like this one forced Bohm to disagree with the thesis of Thomas Kuhn's *The Structure of Scientific Revolutions*.[18] Writing on the history of science, Kuhn had postulated long periods of "normal science" in which the basic paradigm of existing physics is not questioned. Only when scientific progress becomes blocked does a scientific revolution take place, and in it the old paradigm is swept away. By contrast, Bohm believed that all revolutions, whether scientific or political, are partial. The human mind, and human society, work from fixed nonnegotiable positions that tend to be carried over unchanged during a so-called paradigm shift. Even the Marxist Soviet Union he had once supported had only represented a partial revolution. Likewise, quantum theory and relativity might look revolutionary, but at heart they still retained the old order. The transformation of scientific thinking became, in Bohm's mind, inseparable from the wider question of the transformation of consciousness. Bohm was now seeking a new order in physics—a quest that would occupy him for the rest of his scientific life.

While pondering these issues Bohm received news that his father had fallen seriously ill. Samuel Bohm had cancer that had spread throughout his brain to the point that it was inoperable. Realizing that he was soon to die, Samuel expressed a wish to see his son. As a Brazilian citizen, David Bohm had to apply for a U.S. entry visa, which might well be denied. He was unclear as to his legal position since he had never received formal notice that his U.S. citizenship was revoked. In Cardiff Bohm met with an American consular official who confirmed that he was no longer a U.S. citizen. Notification had indeed been mailed to him, it turned out, but to a former address in Brazil.

In the United States Bohm's relatives got on the case, pleading with their congressman for a special dispensation for Bohm's travel. However, Edward Gudeon, the lawyer who eventually had Bohm's U.S. citizenship restored, felt that the whole case was overdramatized. People often appeal to their congressperson in times of crisis, and such officials tend to believe that they have great power, but in actual fact the congressman could have done little to sway the State Department. In the end Bohm was issued a tourist visa and a "waiver of excludability as a former member of the Communist party."[19]

Bohm next worried about the cost of travel, but again the family intervened and sent money for two tickets. It would be Saral's first visit to the United States. None of the Bohm family had met her before, and when she arrived, David's cousin, Irving Bohm, presented her with a bouquet of flowers. To Irving's surprise, David grew quite jealous, telling his cousin, "You don't give flowers to another man's wife."[20]

Each day David and Saral visited Samuel in the hospital. David's father seemed happy that his son had married a Jewish woman, and he enjoyed it when Saral massaged his feet. Even at this stage of his illness, she noted, Samuel Bohm still had an eye for the ladies. Once he knew that his son was happily married, Samuel appeared content. He told them to go off and do what they wanted. Now his only desire was to talk in Hungarian about the old days with a rabbi friend and listen to his niece, Ester, sing him Hungarian songs.

Although it was not normally mentioned, it had long been known that David's father had kept a mistress during the years he lived apart from his wife. Now on his deathbed he mentioned that he knew of a good nurse in Florida who could look after him. The "nurse" arrived, and one day, when Samuel was feeling well enough, he asked his nephew Irving to take him to the bank. There he withdrew money, which he gave to his mistress of several years. In his will he left the rest of his money to his sons, David and Robert.

The other side of the family, the Popkys, suggested that David should visit his mother, who was in a psychiatric hospital in the Pocono Mountains. David and Saral, along with David's brother Bob, took a Greyhound bus to the hospital, which David found depressing. Saral discovered Frieda to be intelligent but particularly aggressive. She argued with them over the issue of having children: It irritated her that none had yet come out of the marriage. Saral made the excuse that they were not yet settled, concealing the true reason—David was worried his mother's mental illness could be passed on genetically to any children.

After the Bohms returned to Bristol, Samuel survived for a further nine months. Bohm did not to attend the funeral, but Saral believed that his father's death moved him—not, however, as much as the death of his mother, when he broke down and cried. It was after the death of his father that he began to speak of wanting a child. Saral consulted her doctor, who told her that because of a medical condition, she would have to spend part of her pregnancy in the hospital. But for Bohm, a separation from Saral was insupportable, and the couple dropped the idea of starting a family.

In April 1957 Bohm presented his hidden variable theory "A proposed explanation of quantum theory in terms of hidden variables at a sub-quantum-mechanical level" at the Ninth Symposium of the Colston Research Society on "Observation and Interpretation." The distinguished audience included Vigier, Popper, Körner, and Leon Rosenfeld. The speaker who followed him was Rosenfeld, who strongly attacked the whole approach referring to such ideas as "dilettantism." A year later another criticism followed, this time from the pen of Werner Heisenberg.

In 1958 a semipopular autobiographical account of Heisenberg's role in the evolution of quantum theory was published. *Physics and Philosophy: The Revolution in Modern Science* appeared as volume 19 of World Perspectives, a series of short books by "the most responsible of contemporary thinkers."[21] After presenting the history of quantum theory, Heisenberg took on the theory's distinguished critics—Einstein, Schrödinger, de Broglie, and Bohm. Since quantum theory leads people away from familiar materialistic views, he argued, it is natural that some of them should want to return to theories closer to classical physics. He thereby portrayed dissenters as clinging to outmoded nineteenth-century views, an old guard standing in the way of new ideas. While this may have been effective rhetoric on Heisenberg's part, it glossed over the deeper problems involved.

When Heisenberg focused his attention on Bohm, it was to make technical objections to the hidden variable approach. He disliked the way Bohm's theory destroyed what he felt to be a basic symmetry of quantum theory, that between position and momentum; in particular, he argued that the wave function should not be taken as "objectively real." Bohm, he said, had introduced a degree of unclarity by treating the wave function in an objective way.

Moreover, Heisenberg maintained, if the hidden variables are truly hidden, then in effect, Bohm was not really able to offer an account of physical process as he claimed to do. In the last analysis his theory could do no better than the quantum theory he was attacking. Bohm tried to get around this problem, Heisenberg noted, by speculating that at some future point his hidden variables might play a significant role in new phenomena. This, joked Heisenberg, reminded him of a man who says that while two times two equals four, he hopes that later it will turn out that two times two equals five, since it would be of great advantage to his finances.

That last remark really rankled Bohm. He originally intended to write a review of Heisenberg's book, but when he realized that most of it would be taken up with refuting the attack, he decided to publish it as an extended article in the *British Journal for the Philosophy of Science,* where it appeared in 1962.[22] Turning the tables on Heisenberg, Bohm thanked him for presenting so clearly the many difficulties inherent in conventional quantum theory.

Far from seeking to return to nineteenth-century ideas, Bohm wrote, he accepts that quantum theory, with its holistic interconnectedness, leads to a radical change in the notion of an "object." Only in the classical limit [of everyday distances and energies] do separately existing things become possible. But even if things no longer have a complete and separate existence, "subjectivism" is not inevitable. It is at this very point, Bohm argued, that conventional quantum theory is inadequate. Indeed, in trying to avoid these problems, he said, Heisenberg himself is guilty of making metaphysical assumptions about "physical actuality," assumptions that are by no means warranted by quantum theory itself. Bohm concluded that the theory must be extended to include a description of quantum actuality. In this context he attacked Bohr's and Heisenberg's argument that the "classical concepts" of space and time are the only ones in which it is possible to describe experimental conditions. In their place he referred to topological notions of "between, inside, outside and neighbourhood." The conceptual language inherent in these ideas, Bohm maintained, lies closer to the way human observations and scientific measurements are made than does that based on Cartesian coordinates.

When it came to a justification of his own hidden variable or causal interpretation, Bohm modified his original position. At Princeton and Brazil Bohm had written to his friends about the great importance of his discovery; it would revolutionize physics, restore causality, and lead people to more rational lives. But while he was writing *Causality and Chance in Modern Physics,* his attitude had begun to change—"many features of the model are implausible and, more generally, . . . the interpretation proposed . . . does not go deep enough," he wrote. "Thus, what seems most likely is that this interpretation is a rather schematic one which simplifies what is basically a very complex process by representing it in terms of the concepts of waves and particles in interaction." And "the author's principal purpose had not been to propose a definitive new theory, but was rather mainly to show, with the aid of a concrete example, that alternative interpretations of the quantum theory were in fact possible. Indeed, the theory in its original form, although completely consistent in a logical way, had many aspects which seemed quite artificial and unsatisfactory. Nevertheless, as artificial as some of these aspects were, it did seem that the theory could serve as a useful starting-point for further developments."

Bohm felt that his own major achievement had been to demonstrate that, contrary to Bohr and others, it was logically possible to create a complete and consistent alternative to quantum theory. In this light the causal interpretation can no longer be criticized since it does not claim to do anything new or exceptional; it is merely a counterexample to the prevailing assumption that quantum theory is the only complete and consistent account of subatomic experiments.

Although Bohm may have adopted a scientifically unassailable position, in another sense it made his theory less interesting: he had retreated from the claim that it reflected something real about the world that held clues to an even deeper account. As it stood in his reply to Heisenberg, it was no more than an abstract counterexample.

His rebuttal did, however, contain hints of ideas that could be taken further. The idea that a new order of space and time could be founded on notions of "between, inside, outside and neighbourhood" was provocative. It had come to Bohm while he was attending a seminar in the Bristol mathematics department. He was struck by the way the basic notions of algebriac topology resonate with quantum theory. In topology a body can be defined as being contained within, or outside, another. Algebraic topology goes one step further than topology itself, doing away with underlying conceptions of space altogether and dealing with pure relationships.

Bohm realized that topological order would fit harmoniously into quantum theory. Bohm wanted not a "new mathematics" in which to express his ideas but an entire new order. The basis of this order was analogous to the algebra of relationships found in algebraic topology. With this insight Bohm began a thread of research he was to pursue at Birkbeck College right up to the end of his life.*

Bohm talked about these new ideas when he visited the Maxwell Society, a group of physics students from London, at their retreat at Richmond Park. Bohm always enjoyed the chance to meet new minds that had not yet settled into conventional tracks of research. In the audience was a young man, Basil Hiley, who immediately responded to the new world Bohm had opened for him. That weekend he decided he must work in the same department as Bohm, and up to the day of Bohm's death he remained his principal scientific collaborator.[23]

While Bohm did not see eye to eye with Thomas Kuhn about the radical nature of scientific revolutions, he did agree that paradigms become embedded tacitly, during a scientist's long apprenticeship. This notion may account for Hiley's openness to Bohm's ideas, since his own apprenticeship had been unorthodox. Hiley had been brought up in the world of the British Raj, the elite rulers of India. Once at a railway station in India, he recalled, his mother had demanded that a private carriage be coupled to the train for herself and her family. The Raj enjoyed first-class service on the railways, however, their education left much to be desired. In Hiley's case, school generally meant a single room in which one teacher attempted to teach fifteen children of widely varying ages. To make matters worse, Hiley's family rarely spent more than six months in any one location.

The result was an early habit of independent study. One of the

* While he was at Bristol, Bohm talked about these new ideas on the BBC *Third Program,* and even today I can well remember listening to that program. I had a passionate interest in science and, living in a Liverpool suburb where not much emphasis was placed on academic learning, much of my education came from listening to talks on the BBC *Third Program* and struggling with science books from the local library. In many ways I felt my own background and upbringing were similar to Bohm's. While at school, I had taught myself the general philosophical principles of quantum theory and relativity, and during the earlier programs, I had listened to more conventional physicists on the topic of quantum theory. Then one night somebody called David Bohm came on the radio, and I was swept away by what he said, by the breadth of his imagination and the appropriateness of his speculations. He carefully explained, to his nonexpert radio audience, some of the basic ideas of topology and why they provided the natural way to deal with quantum theory. From then on, the name David Bohm would always be on my mind, as someone who, after Einstein, had the courage and imagination to engage in big ideas.

significant moments of his youth was being given a copy of Hall and Knight—a classical algebra text used in many British and colonial schools. It opened a door when Hiley discovered that he could work things out for himself without the help of a teacher and even check his results to discover if they were right or wrong.

Back in England Hiley discovered he was older than the other boys in his class. The only way he could complete grammar school (by passing the required "ordinary-level" or O-level examinations) was to drop all the other subjects and concentrate on mathematics, physics, and chemistry. During this period he had another revelation when the math teacher sold him a secondhand copy of James Jeans's *Mysterious Universe* for sixpence. Hiley, who was rather careful with his money at the time, admitted that it was the best sixpence he had ever spent, for Jeans opened up to him a world of scientific imagination.

The university was a disappointment to Hiley; the courses were pedestrian, and no one was interested in the sorts of questions James Jeans wrote about. Once while he was working on his dissertation, he happened to be reading Louis de Broglie's *The Double Solution*.[24] A professor came into the room and took the book out of Hiley's hand, saying, "Don't read this heresy." The event made Hiley even more determined to explore alternative pathways.

It was just as he was winding up his doctoral research and thinking of finding a university position that Hiley heard Bohm talk about his new cosmology. During that weekend Bohm explored ideas on prespace and a process vision of nature in such an open way that he seemed to be inviting his listeners to join him. Other professors set up barriers before their students; Bohm suggested that anyone with a free and open mind could join in this great inquiry. Physics was about understanding nature and not about making calculations.

Somewhat shyly, Hiley spoke to Bohm and learned that he would be moving to the chair of theoretical physics at Birkbeck College, University of London. Hiley himself had been offered positions at Sheffield University and at Birkbeck. That weekend convinced him that he must go to Birkbeck.

For some time Bohm had been looking for a way to leave Bristol. An old friend from Bohm's California days, Eric Burhop, had suggested he would find a sympathetic ear in J. D. Bernal, head of the physics department at Birkbeck. Bernal was a remarkable man. The father of molecular biology, his interests ranged widely, from crystallography and the composition of the Earth's crust to the origins of life. A confirmed Marxist, he was a good friend of Pablo Picasso. He was the sort of open-minded, socially committed scientist who would respond to Bohm's way of thinking.

In the late 1950s Britain began to experience a "brain drain," as its most promising scientists were lured away to the United States with promises of higher salaries and rapid promotion. As a result, applicants for Birkbeck's chair of theoretical physics were few, and the position was offered to Bohm. At the time he viewed Birkbeck College as merely "a step on the way" to a better university.[25] As it turned out, he remained at Birkbeck until his retirement and, despite the offers of other attractive appointments, stayed on as emeritus professor until his death.

CHAPTER 12

Birkbeck: Thought and What May Lie Beyond

IF THE BOHMS were to live in London, why not in the borough of Hampstead, that enclave of artists, writers and intellectuals? Saral soon found an affordable apartment, but Bohm felt the particular location was too noisy. In the end they chose a detached house farther from the center of London, on Gibbs Green in Edgware, the last station on a branch of the Northern Line tube. From his new home Bohm would walk for ten to fifteen minutes to the tube station, where a train took him to Goodge Street Station, close to Birkbeck College. This journey, from home to office, took up to an hour, but Bohm enjoyed the daily solitude it brought him. Even during a period of long rail delays, he was happier to sit on the train and think than take a taxi.

Bohm may have been indifferent to his general surroundings, but for Saral Edgware was a compromise she had to live with. When he was speaking in public and in his earlier correspondence with his women friends, Bohm espoused enlightened ideas about women's freedom and the need for their advancement in the professions. Yet early in their marriage Saral was surprised at how quickly he fell into the traditional male role of demanding a properly run home and meals served at preordained times. While they were living in Israel, Saral had had to rise early to get to work; she took three buses and then faced a long walk. The work itself was physically tiring, particularly in hot weather. Yet when she finally arrived home, Bohm expected his meals to be ready, he made no moves even to wash the dishes, let alone help in the house. At times Saral hit out; if he was so concerned about humanity, she fumed, why couldn't he begin with the person with whom he lived? Since Bohm found it difficult to express anger their quarrels mainly smoldered.

Saral's painting was another contentious issue for it was an area of her life that she desired to preserve for herself. When they were living in Israel, she had painted a woman with a blue face. Her husband became excited at what she had done and explained to her how she should go on to develop the piece. It seemed to her that even while she painted her life was not her own. Indeed, Bohm seemed anxious whenever she was away painting because he liked to have her around the house. Yet if she stopped painting he would question her as to why she was not pursuing her art. On one occasion she blew up, saying she could not cope with all the demands he placed upon her. When he insisted that she continue, she spent the day working on a painting of a meal which she then placed on the dinner table before him.

Saral had continued with lessons in Bristol and now, in London, she attended sculpture classes at St. Martin's School before finding a private instructor in Belsize Park. Again tensions focused around her work and, as Bohm became increasingly involved with the work of Krishnamurti, she seemed to have no more free time. The result was that she took her pictures, brushes, and paints and locked them away in the garage. Part of her life may have ended, and yet, like the child in Edward Albee's *Who's Afraid of Virginia Woolf?* those canvases were ever present between them. From time to time, and in front of her husband, Saral would explain to friends how and why she had given up her painting. Years later, when a friend of the Bohms', the painter and psychiatrist David Shainberg, visited their home, he asked to see her work. She led him to the garage. After looking at the paintings, Shainberg encouraged her to take it up again. As he talked, he noticed that Bohm looked depressed.[1]

In his theoretical world Bohm lived on the edge of the known, yet in his personal life it was difficult for him to relinquish control or accept limits on his freedom. When a relative of Saral's arrived to stay at their home, Saral suggested that the two of them should go into London to attend a concert or visit the theater. For a few days they made plans, but they came to nothing: Bohm expected a hot meal to be served to him when he arrived home from college. Saral simply could not be away from the house at dinnertime. Still, by responding to her husband's demands, Saral exercised an increasing measure of control in the marriage, and more and more Bohm came to rely upon her. Ironically, Bohm had written to Miriam Yevick, before he met Saral, that a woman subjugating herself to a man establishes a claim on his being that leads to her controlling him.

Was Bohm really as self-centered as these incidents suggest? Saral understood that his work and thinking always had to come first. If his mind were to operate properly, he believed, his body had to be provided with certain conditions. Saral was efficient in providing them, preparing his meals and making his travel arrangements. She responded to telephone calls and took care of all his practical needs.

She also helped him in social situations. When he spoke to people, Bohm gave off considerable energy, with an intensity that could inspire. But as he aged, this energy could suddenly vanish, leaving him an empty shell.* At such times Saral supported him with her own energy. Without her, Bohm might not have survived his deepest depressions to live as long as he did.

The result was inevitable; Bohm became increasingly dependent, and following his coronary bypass in 1981, he would no longer travel without her. When he was invited to lecture, he asked that travel funds and accommodation be extended to both of them. He became terrified if he thought Saral was getting ill, since there would be no one to take care of him. This fear evoked similar anxieties he had experienced as a child, when he believed his parents were about to split up. As far as Saral was concerned, Bohm had all the unresolved feelings of a child who had never been loved and cared for. Her role eventually became

* One late evening in 1972, having spent several hours talking to Bohm earlier that day, I passed his open office door and glanced inside. Bohm sat slumped in his chair, his face totally white and drained. While he was only fifty-five at the time, he looked like an old man, so frail that I feared that he had had a heart attack.

that of mother as much as wife. Although this might not have been completely satisfying for her, she did not shrink from it.

In an increasingly feminist society, Saral was well aware that the role she had adopted was anachronistic. She may also have been aware of the criticisms that others made behind her back. Had she sacrificed her life? she later asked herself. During Bohm's lifetime she certainly did not see it so dramatically; she simply realized that if she were not around, Bohm would not be able to continue with his work.

Bohm's admirers—and he had many in his later years—spoke of his exceptional selflessness and his lack of ego. Only a few noted that paradox of selflessness combined with a need for approval and acknowledgement. To put it another way, Bohm's sense of self may have been weak, yet its protection occupied an important place in his life. When engaged deeply in his work, pursuing truth without reservation, Bohm entered states of pure egoless focus that few people ever experience. Yet in other situations he was shy, self-conscious, withdrawn, and insecure; he was well capable of feeling snubbed, rejected, and even betrayed by friends and colleagues. Toward the end of his life, for example, he spoke of the humiliation he had experienced at the hands of Krishnamurti who, in his presence, made cutting jokes about "professors" and did not acknowledge the importance of Bohm's work. If only, Bohm thought, Krishnamurti had been willing to listen to his ideas on the implicate order, that would have been a way in which they could have explored even deeper. Maybe the idea of the implicate order would help people to understand what Krishnamurti was saying. Yet the Indian teacher never seemed to take such things seriously, probably regarding it as only another product of thought. As so while his ego boundary sometimes dissolved into a much greater essence or ground, at other times Bohm's impoverished image of the Self had to be defended.

Despite its internal tensions and mutual dependencies, the Bohms' marriage survived, and Saral supported him until the end. Edgware became his permanent home, and Birkbeck his college for life.

Birkbeck is one of the essentially autonomous colleges associated with the University of London. Its original mandate had been to provide working people with a university education, and so it catered to more mature students, offering lectures in the late afternoon and early evening. Since it took Bohm several hours to get going in the morning, this arrangement was ideal.

Basil Hiley's original intention at Birkbeck had been not so much to work with Bohm as simply to benefit from his presence. Over the years, however, a collaboration slowly developed between them. There was also the department head, J. D. Bernal, who, until his retirement, listened to Bohm with a sympathetic ear and offered him encouragement. Above all Bernal created an atmosphere in which Bohm felt perfectly free to work on whatever attracted him. Although he received offers from other universities, Bohm never seriously considered moving from Birkbeck College.

During his first weeks in London, Bohm looked over the other London colleges. None of their research interested him. One of the senior physicists at Imperial College said that he could not "see the point" of Bohm's work, but then, neither could Bohm see the point of what other physicists were doing.[2]

He met the philosopher Karl Popper, and while they did not always agree, a cordial relationship developed, in which the Bohms visited Popper's home. Popper was sympathetic to Bohm's ideas until the topic of Krishnamurti was raised. To Popper, the Indian's teachings smacked of totalitarianism.

In the mid-1960s Bohm was asked to write a paper for a Festschrift to be published in Popper's honor. When Bohm showed him what he had written, Popper became concerned about a passage: it appeared critical of his work or in some way, he felt, misrepresented him. Popper asked Bohm to remove the offending passage, a request that Bohm found somewhat surprising, having always believed in totally open debate. Unless the passage was omitted, Popper then indicated, Bohm would lose a friend. Bohm acquiesced, but he lost his earlier closeness to the philosopher; they no longer met on a regular basis.*

In terms of the books, papers, and articles he wrote, the lectures he gave, and the visits he made, the Birkbeck years were among

* I have no independent confirmation of this story. Bohm related it to Paavo Pylkkanen, and Saral Bohm also recollected elements of it.[3]

Bohm's most prolific. This was the period in which he formulated his famous implicate order and completed his definitive book on the causal interpretation. Bohm also reached out beyond the physics community to capture a wider audience, writing and speaking about creativity, consciousness, education, dialogue, and healing. He inspired not only physicists, philosophers, psychologists, educators, theologians, musicians, writers, and visual artists but many ordinary people in everyday jobs, who had a deep desire to understand what science was saying about their world and how a positive change could be made in their own lives.

In *Causality and Chance in Modern Physics,* Bohm had acknowledged the limitations of his hidden variable theory. Yet for a time he continued to work on its implications. Although his theory had been published in 1952, its full meaning remained to be unfolded. As I have already pointed out, a new theory is as dense as a great poem and conceals metaphoric levels of meaning within its symbolic relationships. In this sense, Bohm was not fully aware of what he had created in his first hidden variable papers. Thus, during his trips to Paris, he and Vigier had worked at unfolding the underlying meaning of the theory.

It is difficult to reconstruct the history of their intellectual relationship; ideas and attitudes change, and memory is so fallible. Certainly Vigier was one of the few people who had been willing to take the causal interpretation seriously and encourage Bohm in his work.[4] As a colleague of de Broglie, moreover, Vigier felt that he was developing an idea that had originated with the French physicist. But as time went on, their approaches began to diverge.[5] Even today the several groups working on the causal interpretation—or "Bohmian mechanics," as it is becoming known—are fragmenting into different schools.

Bohm's original motivation had been to reinstate causality within the quantum theory and to counteract what he felt to be Bohr's unnecessarily mystifying metaphysics. Bohm's theory demonstrated that the electron is guided along deterministic paths. Yet at the same time he firmly rejected a mechanistic explanation.

He and Vigier explored the idea of a subquantum medium that would account for the chance events produced by quantum theory. In this way it may have seemed possible to produce a theory that predicted slightly different results from those of orthodox quantum theory. Chance is absolute in the conventional theory, but on the basis of a subquantum medium, it may be possible to detect very slight deviations from chance, provided that successive measurements are done very rapidly. Bohm viewed this explanation not as mechanistic but as an aspect of the infinity of levels. While the subquantum medium exerts an effect on quantum events, it too is conditioned by higher levels, and these interweaving levels, in turn, are embedded in yet others.

While Vigier and Bohm remained friends and active colleagues, they increasingly differed in their philosophical approaches. As time went on, Bohm emphasized the role played by the quantum potential. The quantum potential—whose influence depends on its overall form and not on its strength—is totally unlike anything previously postulated in physics, he pointed out. Accordingly, the causal interpretation was far from being a return to classical ideas since it demonstrated the profound difference between old ways of thinking and new quantum phenomena. Later still, in the 1980s, Bohm unfolded a new idea, "active information," to explain the activity of the quantum potential.

Notions like active information, and a potential that depends only on its form and not on its intensity, are radical and somewhat mysterious. There may have been times when Bohm wondered if he should give these ideas their head or step back and encompass them within a more conventional explanation. Bohm would never have avoided "the new," but before his break with Marx, for example, he may have felt that the implications of his new theory were drawing him away from strict dialectical materialism. Vigier, for his part, always cast the theory in materialistic terms.

It is worth placing Bohm's new ideas in an historical perspective. Two hundred years earlier Newton had discovered the solution to the outstanding scientific problem of his day: the moon's orbit around the Earth. Yet he resisted publication of his work because he was troubled by the nature of the force involved. Extending across empty space, he realized, the attraction of gravity is an "action at a distance" rather than the mechanical push and pull familiar to scientists of his day. "Action at a distance" seemed so mysterious that Newton referred to it as an "occult force," in the sense that its nature was hidden. Today gravity, and its companion forces of electromagnetism, are so familiar as to be taken for granted. Yet for Newton, a man who even worked seriously at alchemy, the notion was so shocking that for many years he suppressed it.

To twentieth-century physics, information and the quantum potential appear as bizarre as gravity must have seemed to Newton. In his Marxist days Bohm probably resisted notions that pushed things away from a strict materialistic basis toward mysterious actions and, in particular, to nonlocality.

Bohm argued that the quantum potential "guides" the electron in a nonmechanical way. But how exactly it operates was less clear. Vigier favored explaining the process in terms of some sort of underlying mechanism—a subquantum fluid, perhaps. In his opinion the electron exchanges energy and momentum with this fluid and, in this sense, is pushed along. Bohm went along with Vigier for a time, but eventually he felt that his theory was being forced too far back in the direction of classical physics. The two men simply drifted apart in their philosophies. By the early 1970s his active collaboration with Vigier had ceased, although Vigier continued to interact with students from Bohm's group, such as Peter Holland, inviting them to Paris. Bohm himself was becoming increasingly interested in notions of topology, prespace, and a new order to physics. For Vigier, however, the crisis of the theory lay in the problem of nonlocality.

The question of nonlocality is basic to the original debate between Bohr and Einstein on "independent elements of reality." Einstein argued that if quantum objects are far enough apart, they must be independent; since quantum theory could not account for this fact, then the theory had to be incomplete. Bohr countered that quantum systems are an unanalyzable whole. This same wholeness, present in Bohm's theory, implied nonlocality.

Nonlocality, as we have seen, means that distant objects are correlated in ways that classical physics can never explain. *Correlation* does not mean mere synchronization, like two digital watches that continue to tell the same time even when they are far apart. Rather, it is a more active form of correlation that keeps the two particles corelated. Nor does nonlocality imply the sort of synchronization that exists between clocks linked by a radio signal. Nonlocality, according to conventional quantum theory, has nothing to do with interactions and signals; it is instantaneous and appears to transcend the limits of space and time.

Quantum nonlocality cannot therefore be explained on the basis of any conventional field or force; it does not involve signals being sent from one particle to the other. The concept is so difficult to explain precisely because our very language, like our everyday concept of space, is based on fundamental notions of location and separability. Talking about nonlocality demands the creation of a new language. When nonlocality was found lurking in the heart of Bohm's hidden variable or causal interpretation, it came, according to Vigier, as a great shock to both men.

From Vigier's perspective, Bohm had created his original theory in a bid to restore causality to the heart of physics. Yet now it appeared that nonlocality had reemerged to deny causality. He argued with Bohm about the meaning of this nonlocality and later recalled thinking up all sorts of causal paradoxes that follow from it—for example, a person travels back in time to meet his father and so prevents his own conception.[6]

Thus, in Vigier's opinion, nonlocality subverted their entire program. So disturbed was he that he abandoned the research altogether and moved into the field of cosmology. He believed that Bohm felt the same way about things, and that it was for this reason that Bohm gave up physics and turned to the teachings of Krishnamurti. Only later, when it was shown that causality is not violated by nonlocality, did Vigier return to the de Broglie–Bohm theory.

It should be added that this is Vigier's own interpretation of the events. Bohm may well have been puzzled by nonlocality, and he once suggested to Basil Hiley that an absolute nonlocality may not be possible. But neither Hiley nor Philippidis, a student who later worked with Bohm, believed Bohm's reaction had ever been so extreme as to cause him to abandon his causal interpretation. Indeed, his hidden variable approach is essentially a local theory that gives what appear to be nonlocal results!

Why then did Bohm give up his hidden variable work? As far as Hiley was concerned, Bohm simply lost interest in his 1952 paper in favor of his work on topology, prespace, and a new order to physics. In fact, he did not return to the theory until the early 1970s, when Christopher Philippidis began an independent investigation of hidden variables.

Bohm may have put aside hidden variables, but as far as his colleagues and students were concerned, his passion for physics continued unabated. His lectures and seminars were inspiring, and he radiated ideas when he spoke to his doctoral students. Yet in private, Bohm had periods of dark despondency in which he questioned the value of his work and wondered if he should abandon it altogether. His general discouragement, which was particularly debilitating during his first years in London, was relieved only by his relationship with Krishnamurti. The Indian teacher pointed to a direct awareness of a universal ground, the same ground Bohm that was trying to describe in his physics. As they talked, Bohm's focused energy enabled him to enter into and remain within the "untalkable," at the same time pushing the Indian teacher to clarify and expand his teachings.

In his public talks and private conversations, Krishnamurti would address the nature of thought and the transformation of consciousness. Thought, for Krishnamurti, is a constant activity of the physical brain. (And by "thought," Krishnamurti included feeling and, indeed, all the content of consciousness.) This activity, he explained, operates in response to memory, a record of the past or a sort of internal image. He admitted that thought and the response to memory can be very useful in its proper place—practical matters such as driving a car, doing mathematics, or carrying out scientific research—but when it comes to psychological matters it gives rise to all manner of conflicts and confusions. To take a daily example, a husband may not respond so much to a remark made by his wife, but rather to his whole mental image of his wife, his memory of past remarks, and the complex emotional associations involved. "Thought is a movement in time," Krishnamurti said, and the husband is trapped within by this psychological dimension of time, unable to enter into the actuality of the moment.

This problem arises, in part, because the thinker assumes that he is quite separate from the thought, that he is in control of thought and able to observe it objectively or influence its outcome. But, as

Krishnamurti so often emphasized, "the observer is the observed, the thinker is the thought." The more one attempts to control one's thought, for example, by condemning one's angry reaction, the more an inner conflict is established that feeds the general process of thought.

For Bohm, this insight had strong resonances with his own thinking and with the quantum mechanical inseparability of observer and observed. Indeed, several years later, he and the environmental photographer, Mark Edwards, wrote a book together, *Changing Consciousness: Exploring the Hidden Source of the Social, Political and Environmental Crises Facing our World,*[7] which traced our present environmental and social troubles to the operation of thought.

Many religious and philosophical systems had associated conflict and suffering with the operation of thought. They had also offered practices for bringing thought into proper order. But, for Krishnamurti, any practice, technique, approach, or discipline perpetuates the overall situation by separating observer and observed. There is, he taught, no way in which thought can be controlled or made to cease for there is no agent who exists apart from the process itself. Rather, the entire movement has to cease so that something qualitatively different can come into operation. Krishnamurti spoke of this as "dying to thought," suggesting that when this takes place the physical brain no longer supports or gives energy to the movement of thought and thereby becomes totally silent. It is from within this silence that the unconditioned operates.

This was the "transformation of consciousness" that Krishnamurti referred to. For him, it could not be a partial thing, or something that operated in a temporary manner. If the transformation occurred then it would be total and permanent. At times Krishnamurti would refer to the unconditioned as "the intelligence." But before his dialogues with Bohm, Krishnamurti had rarely spoken in a specific way about the nature of this "intelligence." Now Bohm was asking him if its operation brought about a physical mutation of the brain's structure. Krishnamurti agreed that this could take place and that it was possible for brain cells to regenerate.

There were times when, in dialogue with Bohm, Krishnamurti

became so moved that he was forced to leave the room, saying that he had never openly discussed such matters before.[8] On one occasion Krishnamurti said that there was no thought present within the room. Bohm believed that there was a little thought moving within his brain, but Krishnamurti told him that whatever thought arose was not held on to but was allowed to die away. On another occasion, as they entered very deeply into the idea of universal mind, the excitement in the atmosphere became palpable. Krishnamurti held on to Bohm, saying, "What did you feel?" "I didn't feel anything," Bohm replied. "There was just great clarity." "That's it!" Krishnamurti replied.

Krishnamurti claimed to have undergone that transformation of consciousness that occurs when the mind "dies to thought." As a result, he taught, his physical brain was totally silent and his perception of truth was immediate and direct. But could others enter into this state? those who had followed his teachings asked. Bohm himself believed that such a transformation could occur if sufficient energy was generated when a person was in dialogue with Krishnamurti himself. After all, he believed, the mind is a nonlocal. If only ten people underwent this profound change, he once told me, they would act as a nucleus to change the whole of society.[9]

His discussions with the Indian teacher were so important to Bohm that he and Saral traveled each year to Saanen, Switzerland, where Krishnamurti gave public talks. It was during this period that Bohm talked to him about his physics. Krishnamurti did not think much of science, or for that matter of music, art, philosophy, or literature. But he nonetheless advised Bohm to "try to begin from the unknown." For Bohm, the crucial issue was how to dissolve the rigid compartments of knowledge and allow something new and creative to operate.

Thought is the tool whereby science proceeds, yet thought also controls the thinker. The deepest scientific insights come about, Bohm believed, only when the mind reaches a state of such intense energy that habitual patterns of thought are dissolved. The creative moment of an Archimedes, Newton, or Einstein involved the same transformation of consciousness of which Krishnamurti spoke. Yet enlightenment, for these scientists, occurred only for a brief moment and did not embrace the whole of their lives. They rapidly fell back into an ordinary state of consciousness that Bohm described as rigid and brittle. Krishnamurti, by contrast, spoke of something radical and total.

Krishnamurti demanded that thought be pushed to its limits, to the point where it gave way to something else. Bohm, however, believed that thought still had a significant role to play. When it was "brought into order," it could be a valuable tool. Thinking was the center of his life; how could he carry out physics without it? Such questions continued to trouble Bohm.

Their first visit to Saanen was an idyll for Saral since, after Krishnamurti's daily talk, she and Bohm would walk in the mountains alone and picnic high above the town. A combination of environmental factors made that particular spring unique in its abundance of alpine flowers, and the couple were particularly happy together.[10]

At that time Bohm was not singled out for any special attention by the other listeners to Krishnamurti's talks. Then, during the next season, Krishnamurti referred to Bohm by name. Everyone's interest was now aroused, for clearly Bohm was someone whom the Indian teacher considered important. Now the Bohms began to receive invitations to dinner from the other listeners. From being a couple on their own, they—or at least David Bohm—became the center of attention.

Krishnamurti had occasionally told young people that celibacy was significant, indicating that it encouraged the generation of great energy and intensity that could lead to psychological transformation. Krishnamurti seems to have raised the matter with Bohm as well, and the physicist believed that the Indian teacher led a celibate life.

It was during this period that Krishnamurti's interaction with Bohm became more sustained and intense—so much so that one day Bohm announced to Saral that Krishnamurti had suggested the two men should live and travel together, exploring questions on a continuous basis. Bohm was moved by this request, although as it turned out, it was something Krishnamurti proposed to others, sometimes on the basis of only a short acquaintance.[11] Saral, extremely surprised, told her husband that if this was what he truly wanted, they would try to work something out.

In the end Bohm chose not to live and travel with Krishnamurti. He did, however, become deeply involved with a new educational experiment. When Krishnamurti spoke of the transformation of consciousness that comes about when the mind "dies to thought," he meant that thought is the mechanical, conditioned response to memory. Would it be possible to bring up young people without much of this weight of conditioning? he asked. For a number of years Krishnamurti resisted requests to open a school in Europe. Earlier schools had been associated with his name in India and the United States, but in recent years nothing in the West reflected his philosophy of education-without-conditioning. Only after his encounters with David Bohm and with Dorothy Simmons (who, along with her husband, Montague, had been running a school for high-IQ delinquents at Royston, near Cambridge) did he agree that such a venture was worth pursuing. He asked the Bohms, the Simmonses, and Mary Cadogan to look into the idea.

The year was 1967, and Bohm was about to attend a conference in Italy when Krishnamurti asked him to lunch and indicated that he wanted the physicist to "run the school." This request came as a great surprise, particularly since Bohm had little administrative experience and was employed as a full-time professor of physics. While Bohm was in Italy, Saral met with the planning committee; and, while speaking at one of their meetings, she was cut short by Krishnamurti with "She's not in it."[12] It was a hurt Saral could never forgive nor, for that matter, understand. It seemed to her that her only value in Krishnamurti's eyes was as Bohm's wife. Her feelings toward him had always been mixed, but from then on he became an increasing irritant in the marriage.

Back from Italy, Bohm explained that he could not take on the responsibility of the school.[13] In any event, Dorothy Simmons, who already had experience running a school, was to take on the day-to-day responsibility. Probably what Krishnamurti had in mind for Bohm was more the role of a resident consultant or some similarly significant position.

With plans for the school firmly established, Krishnamurti began to look around for suitable property. When Bohm was asked to act as a trustee, he became worried that he could be legally responsible for any debts the school incurred if property was purchased with inadequate fund raising. As expected, he entered a state of anxiety, not wanting to refuse Krishnamurti but exhibiting his lifelong insecurity about money. His hesitation in making a definite decision may well have been, for Krishnamurti, further evidence of a lack of commitment. Eventually property was purchased at Brockwood Park, close to the village of Bramdean, near Winchester, and Bohm then assumed the position of trustee.

After the Brockwood Park school opened, it became Krishnamurti's base in England for several months each year. Bohm finally accepted the roles of both school trustee and trustee of the Krishnamurti Foundation. Throwing himself heart and soul into the life of the school, he drove with Saral to Brockwood each Friday, only returning late Sunday evening. All the passion that he had once given to physics was now being poured into this educational experiment. He talked regularly with the staff and students, trying to resolve difficulties as they arose. He conveyed to them something of his own search for fundamental truth. As he spoke, he made his audience feel as his equals, coparticipants in the same search. And when he dialogued with Krishnamurti, he acted as stand-in for all of them, articulating the questions they would like to have asked.

Following Krishnamurti's lead, the Bohms become vegetarian. Also like the Indian teacher, Bohm took up meditation. But a word of qualification is needed here: By meditation, Krishnamurti meant something very different from the mental exercise practiced in Eastern religions and philosophies. Such systems came in for scathing attack from Krishnamurti: repeating mantras and following gurus were, he said, particularly stupid ways of wasting time. Any system, effort, belief, or intention directed toward a goal perpetuates the artificial duality that fuels the continued operation of thought.

Meditation, for Krishnamurti, was instead an act of alert attention. He likened it to coming upon a tiger in the forest or a snake on one's path—a sense of total alertness and awareness. In meditation one makes no effort to change, improve, transform, or reach a goal. One simply perceives. Anger, fear, and love are also opportunities for meditation—allowing an awareness of the total nature and structure of what is occurring at each instant. Yet as soon as one assesses such an emotional state, judging anger to be inappropriate or love a virtue, then one falls into the trap of separating observer from observed and the thinker from thought.

Bohm had learned from Hegel how thought attempts to hold fast and discover security. In the very act of its arrest, thought discovers its opposite, its antithesis, and in so doing, it begins to move again. No effort, no desire, no goal will ever cause that movement to cease. For in every desire lies the creation of an illusion—the separation of the thinker, the "I," from that which it seeks. In the very desire to arrest thought, the ceaseless movement is born again and sustained. Only something of a totally different quality can bring about its death. This, for Krishnamurti, was the heart of meditation.

Bohm was aware of an intense, attentive watchfulness in Krishnamurti. He noted in particular the attention Krishnamurti gave to the movements of the eyes. It was important to pay attention to these movements during meditation, the Indian teacher once told Bohm. Possibly he meant that the movement of the eyes indicates the movement of thought reflected in the scanning movement of attention.[14]

Bohm set aside a quiet period each day and, while walking, he would watch the movement of his thought. Anger, fear, and thought are, he suggested, like the weather. They are physical, electrochemical movements within the physical brain. Just as we watch the weather without thinking to change it, so too we can watch our thoughts and feelings without getting caught up in their contents. How much Bohm was able to achieve and sustain such a nonjudgmental awareness was not at all clear to those who knew him. Bohm often appeared to be so caught up in anxieties that they permeated his entire being.

The exploration of Krishnamurti's teachings had become of great importance to Bohm. Saral was also impressed when she first heard the Indian teacher speak and was enthusiastic about the idea of a school. Yet as time went on she became increasingly concerned with the way Brockwood was dominating her life. Each Friday they would drive down to the school; she could no longer set aside the weekend to meet with friends. Even if the couple were to take a vacation together, it had to be at Saanen near Krishnamurti. It was almost as if the Indian teacher had become the third party in their marriage, a triangle in which her own contributions were not being acknowledged. Exclusion was painful for her. As a child, she had been given less attention than her brother, who was frequently ill. Behind her cheerful, optimistic exterior lay distant hurts and insecurities.

Things became more even difficult when Saral expressed misgivings about the changes she began to notice in the school. During their first years with Krishnamurti, everything had been open, but now she felt that many of those who surrounded Krishnamurti were placing an increasing emphasis upon the Indian teacher's image. For example, elaborate means were being used to record and preserve each of his talks and conversations. As far as she could see, he was being turned into a guru. Bohm countered by saying it was no longer sufficient that Krishnamurti should simply talk but that his teachings had to be more widely disseminated. Saral hit back, saying that people at Brockwood had begun to make compromises which went against the spirit of Krishnamurti's teachings. Bohm retorted that Saral saw things in black and white and was unable to make allowances for people. As time went on she found it increasingly difficult to speak critically to her husband about Krishnamurti or the school. As with his earlier devotion to Marxism and Soviet Russia, she felt that Bohm was unwilling or unable to look at what was happening.

For Bohm, his mutual exploration with Krishnamurti remained the most significant encounter of his life. The Indian claimed that he was not reproducing a traditional teaching, that at the very moment he spoke he saw truth afresh, as if for the first time. Now Bohm had become a coparticipant in the process, and together they were bringing truth into manifestation.

Some of those around Krishnamurti resented their closeness. They noticed that the Indian teacher's imagery and vocabulary were changing. Bohm, they felt, was overly intellectual and dry, his influence on Krishnamurti was negative, and he was removing poetry and mystery from the teachings.

The tensions came to a head when it was decided to publish the dialogues he and Krishnamurti had held in the early 1970s. These dialogues were so popular that copies of the audiotapes were constantly being borrowed from the school. Edited into book form, the conversations would provide an important avenue whereby a new generation could learn about Krishnamurti.

The task of preparing the manuscript fell to George and Cornelia Wingfield Digby. (George was keeper of textiles at the Victoria and Albert Museum and a recognized authority on William Blake.) But when their completed manuscript was shown to the publication committee, one of its members felt that it did a disservice to Krishnamurti. Not only had his language changed, but a word count showed that Bohm uttered more than the Indian teacher himself. Readers would believe that Bohm was instructing Krishnamurti. Clearly the manuscript could not be published as it was.[15]

The committee sought Krishnamurti's advice. With fine political acumen he threw the ball back into their court, telling them that publication decisions were their province, not his. In the end, with the threat of resignation of one of the committee members hanging over them, publication was suspended.

Bohm heard rumors about the aborted project, but nothing was said to him directly. One day during lunch, a member of the foundation turned to Krishnamurti and said, "You have something to tell David, don't you?"[16] Krishnamurti appeared embarrassed, and when the others left, he indicated that the committee had decided not to publish the book. Only certain dialogues would be included in another publication. No explanation was offered, and Bohm assumed that it had something to do with the quality of the recordings. Only later did he learn the real reason. For many years this incident nagged at him like an aching tooth. Why had Krishnamurti not been direct with him? Why had he not explained the true reason?[17]

Despite such occasional doubts, however, Krishnamurti's teachings filled Bohm's early years in London, almost to the point of displacing his interest in physics. In his dialectical materialist period, physics had been infused with hope, but now he felt it could bring no fundamental change to the human race. Knowledge had little to do with the transformation of human consciousness.

Paradoxically, while Bohm harbored these doubts about the value of physics, his students and colleagues at Birkbeck College knew nothing of them. Whenever he spoke about physics, his energy mounted and his mind generated new ideas. Basil Hiley, for one, found it exhilarating to work with Bohm, who also inspired a number of exceptional students. Yet despite his close daily contact with Bohm,

Hiley never suspected the extent of the physicist's involvement with Krishnamurti; Bohm appeared able to compartmentalize the various aspects of his life.

In addition to his close involvement with Krishnamurti, Bohm retained an interest in the question of a new order in physics. Soon he got an additional stimulation from an American, Charles Biederman, who was not a physicist but an artist.

Born in 1906, Biederman's abstract paintings and geometric reliefs, along with his books, *The New Cézanne* and *Art as the Evolution of Visual Knowledge,*[18] had stimulated the constructivist movement in Canada and the United States. In England Biederman was an important influence on painters like Victor Passmore and Anthony Hill. Hill, feeling that Bohm's approach would interest Biederman, sent him a copy of *Causality and Chance in Modern Physics*. Biederman did enjoy the book, and on March 6, 1960, he wrote a note of appreciation to Bohm.[19]

Bohm acknowledged the note, remarking that he felt a link existed between science and art. Thereupon Biederman invited him to contribute to *The Structurist,* a magazine "devoted to that contemporary art that followed out of the Dutch school of de Stijl." Bohm replied that he felt unqualified to write about art, particularly as, until quite recently, his "reactions to modern art were almost entirely negative."[20]

"You write that you do not feel qualified to deal with the relations between physics and art," Biederman replied. "Do you know of any single soul who is? I do not. Whether physicists or artists, until some individuals in both fields try their best to deal with this problem either they must be humble and appear reductionists, or either be a qualified person in either field."[21]

The result was not a magazine article about art and science but a highly active correspondence that lasted for several years and extended to some four thousand pages. At times Bohm was so excited about the interaction that as he was taking his letter to the mailbox, a new idea would occur to him, and he would dash back home to write an additional ten to fifteen pages.

It was a true meeting of minds, each rejecting establishment thought and proposing radically new theories. At a time when the international art world, and influential critics like Clement Greenberg, were taking New York as their center, Biederman had exiled himself to Red Wing, Minnesota, poured scorn on the current heroes of modern art, and looked back to the European tradition. Combined with his uncompromising and somewhat irascible nature, his work was inevitably neglected.

Biederman felt himself heir to the work of the American scientist and linguist Alfred Habdank Skarbek Korzybski, whose seminar he had attended in 1938. Korzybski was best known for his 1933 book, *Science and Sanity: An Introduction to Non-Aristotelian Systems and General Semantics*.[22] The aim of his "general semantics" was to increase the human capacity to transmit ideas from generation to generation. Biederman applied similar evolutionary notions to the visual sense, noting transformations that had taken place in art from the naturalist Courbet, through the impressionist Monet, to that pinnacle of twentieth-century art Cézanne. Just as Bohm was looking for a new order in physics, Biederman was seeking a new order in art. His ideas became an inspiration and stimulation for Bohm.

Earlier art, according to Biederman, had been mimetic, producing a surface illusion of the natural world. Monet had announced the end of this process. Rather than attempt to reproduce a motif he struggled to emulate the very processes of light on the canvas. In this way he freed painting—and in particular, color—from the attempt to capture appearance. Cézanne had gone further, liberating both color and form with multiple rectangular planes painted in advancing and receding colors. These painters were searching for what Biederman called "the structural process level of reality." This evolution was continued by Mondrian, who, according to Biederman, had then fallen into the trap of idealism and mysticism.

The torch was taken up by Biederman himself, who, in a series of constructions, believed he had freed painting from the two-dimensional—and brought it not into sculpture, which was purely mimetic, but into painting by means of a series of planes and rectangles arranged according to a controlled asymmetry. Most important, rejecting mimesis led not to abstraction but to the manifestation of nature's underlying structural processes. Biederman wanted his work to be viewed in a natural environment so that, like Monet's "fugitive sensations," the colors and cast shadows of his structures would change during the day. Biederman claimed as the goal of his art nothing less than the transformation of human consciousness.

Bohm, who himself sought a radical new order, was excited by Biederman's ideas, "You mention . . . that Cézanne saw the universe as a pulsation of color. You remind me of some of my ideas on the wave-particle character of matter. We must explain why the same thing, e.g. an electron, can in some conditions act as a wave and in others like a particle. Now, I propose to do it by giving up the notion that an electron is a permanently existing particle (like a microscopic copy of a billiard ball). Instead, let us consider a pulsating model of the electron. We should begin with a continuous background field extending over the whole cosmos."[23] Bohm went on to explain his concept of the electron as a field that alternately collapses inward and extends out through the universe.

Biederman introduced Bohm to the underlying order of impressionist paintings, in which each brushstroke combines to produce an overall effect. The value of a particular gesture or color is determined not individually or locally but by the overall context in which it is placed. This context-dependence struck Bohm as similar to context-dependence within quantum theory—the way a quantum system is defined by the context of a measurement or observation. A mathematical description of an impressionist painting, Bohm speculated, would be similar to the new description of quantum theory that he was seeking; one in which form and content cohered.

Monet had created space out of color; Bohm was attempting to derive space from something that was not space. In Monet the value of each element is determined by the order of the whole; likewise in quantum theory the basic descriptive element (a vector in Hilbert space) depends on the whole. Again, just as the value of individual elements in a Monet painting are determined by all the others, so in Bohm's new approach the elementary particles are determined by the whole.

Bohm brought these ideas back into Birkbeck College, where he discussed new notions of order—and in particular, the artistic explorations of Cézanne—with his colleague Hiley. Ideas of the intimate relationship between the parts and the whole, the whole and the parts, were present both in Cézanne's paintings and in Bohm's earlier work on topology and cohomology (which deals in the concepts of boundaries rather than in Cartesian concepts of coordinates). Bohm was also excited that movement and process could be captured in painting. One picture in particular had struck him: It was by the seventeenth-century Dutch painter Salomon van Ruysdael and depicted the flowing movement of a waterfall. It caused Bohm to speculate about the possibility of developing a static mathematics that would represent movement and flow. So great was Bohm's excitement that Hiley was persuaded to visit the Tate Gallery and see some of Biederman's work.

Picking up on an insight of Biederman, Bohm also developed the notion that order was a collection of significant differences, and of the differences of these differences. It was the language Bohm later brought to our book *Science, Order and Creativity*. The key to order, and to information, is difference and similarity—more specifically, collections of different similarities and similar differences.

Some readers may have assumed that Bohm picked up the idea of information from Gregory Bateson, who spoke of "information" as being "the difference that makes a difference." But Bohm and Biederman developed these concepts quite independently. On February 24, 1961, for example, Bohm wrote:

> Thus identity which is at first defined as no kind of difference at all. I.e., absolutely different from every kind of difference. There now seems to be a special kind of difference viz, difference from every other kind of difference. This is clearly a counter diction and one that can be in no way avoided. Likewise the very use of the difference is a counterdiction. Any two things are by definition different and each example of difference is different from every other.

Together Bohm and Biederman also explored the psychological dimension of art: the nature of perception and visual conditioning. Inevitably this exploration took them into issues of consciousness. The ego, they argued, is not a fixed thing but a process—"the ego process."

> Here on the last day of 1962 I would like to say the last word on the development of the ego process. At least let us follow AK's

> [Alfred Korzybski's] injunction to be silent about it for a while. Finally I think the general features of the ego process have changed only very slowly, not basically in the past 10,000 years or more. The features are Identification, Fragmentation; Confusion; Conflict; Escape; Desire for the Pleasant, Fear and dislike of the Unpleasant; Desire to continue the pleasant, to fulfill satisfactory possibilities and to move to a [word unclear] . . . the opposite in a secure fashion; Loneliness and sorrow which follow upon frustration and disappointment in the above in particular leading to fear of death, illness, misfortune, disgrace, defeat, inadequacy, etc.
>
> I think it is very clear that primitive tribes and modern society show the above features.[24]

Bohm's letters to Biederman are starbursts of ideas. Written in a white heat, Bohm poured ideas down on paper, making way for yet more ideas that tumbled from his brain. The result is not easy reading, but even one letter taken at random—written in August 1962—shows the passion of his mind.

> I think the crisis of consciousness in its totality in the form of the ego process is basically similar to that in art, for consciousness . . . is based on the image, the symbol, word, idea, thought . . . each of these is really a fragment, a partial reflection and projection. Our confusion is that we imagine that they are all integrated into a whole. As a result, we see one process of reality reflected as many. One Ego process reflected as observer and what is observed; one physical process reflected in observed apparatus and observed electron and so on. Consciousness is a kind of internal imitation of the form of images, words, feelings, etc and it's based on the active response of memory, i.e. the playback of the memory scratch. On the other hand where this in general goes beyond recognition in the response of memory and this can be creative. Awareness is the perception of concrete reality from moment to movement.

Bohm went on to discuss the nature of mimesis in art and related it to the process of consciousness.

> It seems to me that mimesis in art resembles the actual response of memory which imitates reality as it was, but of course only in a

very superficial way. Our problem is to perceive concrete reality, as it actually is from moment to moment and not on imitations based on remembered images, symbols, literary allusions, emotions etc. Your considerations of the problem of mimesis in visual art may be of crucial importance, perhaps of far more general significance than is generally realized. It may be that the ego process itself is coming to a kind of mimetic crisis. The ego process is based on mimesis through the response of memory but a new kind of broad and deep awareness exists in some way in every age in the past as our society resolves a crisis. In this total revolution the scientist and artist will probably play a key part with what is the role of words. The identification meaning of a word is to evoke the playback of memory scratches. Words . . . point like signs to some process that is concrete reality . . . the role of words is necessary as a continuing reference point in order to start a discussion but in addition there is the operative role whereby they direct us to concrete actuality and then pull away like dead leaves.

The mathematical symbols have similar properties. They are symbols like x, y, etc. They stand for unknown or poorly known entities. In the beginning they are so poorly defined that all they do is establish distinctions or differences between entities without saying how they are different. Then there are the operative symbols like + − × etc. These are like the operative words that direct us to do something with the symbols. The real meaning of entity symbols is realized only through the operative symbols, the operative symbols that lead us into the actual thought.

Very generally words and symbols etc. are abstract reflections of concrete reality. The man I see in a mirror is an abstract reflection of me, but the process of reflection involving the mirror and light rays is concrete. Similarly I am now thinking of an entity called Biederman, or labeled Biederman. My thought is a concrete process of reflection but its content is an abstracted reflection and the real Biederman process which is known better by you than by me.

And as to process, order and structure;

It is important for me to state my ideas on process once again because they have evolved since the last letter was written. The main change is that I now admit, and agree with you, that process exists even in

> the so called static object such as a picture. Of course I realize that no object is really static, but rather it persists in a certain form as a consequence of an inner movement that balance each other, e.g. atomic movements, pulsation etc. But what must be added is that, for this reason, the persistent aspects of the form of the thing can and indeed must reflect process, the example of a sea shell is extremely cogent.

Their intense correspondence lasted for almost ten years. At one point Bohm started a book called *The Chemistry of Thought,* based on this interaction. But the manuscript never saw the light of day.

Within this fertile relationship, tensions were nonetheless present. Bohm found Biederman overly dismissive of the possibility that there could be creativity in paintings other than his own. When Bohm expressed his somewhat mystical experiences while looking at a Georges Rouault painting, Biederman dismissed his reactions as mere confusion. Writing about *The Old Clown* in the Edward G. Robinson collection, Bohm had said;

> At first, it seemed to be rather a mixed up set of patches of color. But gradually, it began to take shape. In particular two patches struck my eye, one in the face of the clown and another outside him which seemed to complement the first. My eye began to move back and forth from one patch and the other, a pulsation was established, and suddenly it ceased, to give way to a remarkable new steady vision which I can best describe as seen in a new dimension. It was not so much that the clown became visible in three dimensions, this was true but only a minor point. The main point is that there seemed to be a flow or a current in which the whole being of the clown poured outward to reveal itself, all his feelings, thoughts and emotions etc., and a counter-flow in which the outside (including the viewer) was drawn into him, to emerge again in the outward flow. It was a very striking experience for me, one that I shall always remember. Whether the artist intended the picture to be seen in this way, I don't know of course. I would be interested in knowing whether it struck anyone else in that way.[25]

Biederman replied, "It seems to me your reaction to the Rouault picture was the result of your experience as a physicist. As I read this paragraph I almost expect you to say something about a breakthrough about the quantum barricade. Am I being absurd?"[26] Bohm was hurt by Biederman's dismissal of what he considered a significant experience, one that he was to recount on several other occasions.

The final split grew over the two men they referred to as "JK and AK." Biederman was a proselytizer for AK, Alfred Korzybski. When Bohm read *Science and Sanity,* he was struck by the way "AK's teachings complemented those of JK" (Jiddu Krishnamurti). The more Bohm referred to "JK," however, the more misgivings Biederman expressed.

In one of his letters, Bohm explained his ambiguous attitude toward science and, at the same time, hinted at his motivation for exploring Krishnamurti's teachings.

> So this is where you and I disagree. I see that you are deeply interested in art as a mode of creativity, and to this I can only say "More power to you." But for me, science is deeply bound up with all my conflicts. You may advise me "Grow continually in creativity, and thus get out of conflict." But for me, to start to do this scientific work is to start to engage in conflict. To realize this consciously is not enough to stop the conflict as soon as all my attention is centered on what I am doing, the unconscious goes into operation, creating conflicting demands, distorting thought, etc. Therefore, *for me,* creativity in science is not an answer to conflict. I must understand conflict *directly* in its psychological sense. But as I start to do this, I begin to see that while the whole field of science has its interest and value, it is rather petty to take it as the central issue in life. I can even see that an improved scientific understanding of the world will reflect back into saner thought, and therefore to an improved means of dealing with psychological problems. For not only is our thought conditioned by the unconscious. But the activities of the unconscious, as well as its structure, are conditioned by our ways of thinking about fundamental questions, such as space, time, matter, existence, etc. But I cannot avoid seeing that science is very limited in this regard.[27]

Biederman believed that art aids the transformation of consciousness, which is conditioned by a particular tactile quality. Through art it should be possible to transform the human race, freeing it from certain emotional problems. Krishnamurti, however, did not think much of art, and Bohm's reports of this view irritated Biederman to the point of a final break.

Even after their correspondence ceased, however, Bohm continued to develop the work he had explored with Biederman and Krishnamurti, bringing it back into his physics. He moved ever deeper, seeking an underlying structure for quantum physics. He had already rejected Cartesian coordinates, with their assumption of an underlying continuum to space, in favor of topological relationships. "Inside," "outside," and "neighbor" seemed more appropriate to the quantum world than dimensionless points.

Yet even these concepts are based on the notion of an underlying space. Why not abandon space altogether and retain only pure relationship? After all, algebra had been defined as "the relationship of relationships." Bohm's goal was an algebra of relationship. It would apply at the level of quantum phenomena, while in the large scale it would average out into the familiar properties of space, time, and matter. Maybe it would also be possible to get rid of those infinities that plagued quantum field theory. Over the next years Bohm moved far beyond his causal interpretation into something more fundamental and more radical.

In Birkbeck's mathematics department Roger Penrose was also trying to get rid of the continuum in physics. First he used what he called spinor networks, but soon he developed a new mathematical object, a generalization of the spinor called a twistor.*

Until the mathematics department moved from Birkbeck's Malet Street building, Penrose and Bohm had run research seminars together. His health improved, Basil Hiley also attended these seminars. While Hiley at first felt out of his depth, he was fascinated by the discussion. Bohm always seemed to be walking around with a book

* Penrose pursued twistor algebras throughout the 1970s. During the 1980s he reintroduced the notion of a continuum into his formalism. Accounts of Penrose's work are fairly technical. The only popular account I know of is in my *Superstrings: The Search for the Theory of Everything*. (At the time I wrote the book, I had intended to call it *Superstrings and Twistors* but was persuaded by the publisher to change the title.

on harmonic integrals in his hand. At the seminars he would go though some part of the book, showing in a very simple way how its mathematical theorems could be related to new ideas on order in physics.[28] As the groundwork for new ideas was being laid, Hiley was moving from an observer to a participator in the process.

CHAPTER 13
Language and Perception

IN JULY 1968 Bohm organized an informal colloquium at Cambridge on the meaning and interpretation of quantum theory, along with the physicist Ted Bastin. The participants included several of Bohm's former students and colleagues, such as Mario Bunge, Yakir Aharonov, and Jeffrey Bub.

This meeting gave Bohm the opportunity to talk to a wider audience about his new ideas on order. Basil Hiley had lately been meeting with a mathematician, R. H. Atkin, from the University of Essex, and now felt he knew sufficient mathematics to make a contribution.[1]

A result of the colloquium was that Bohm and Hiley, along with the mathematician Alan Stewart, began an active collaboration. The first fruit, "On a New Mode of Description in Physics," published in the *International Journal of Theoretical Physics,* applied cohomology theory (a branch of algebraic topology) to physics. At the time, the paper seemed quite speculative, although today the idea of using cohomology in theoretical physics is widely accepted.

Bohm and Hiley set out to express the laws of nature in purely cohomological, or relational, terms without recourse to the usual differential equations of physics or even an underlying space-time. In the nineteenth century Maxwell had formulated his famous theory of electromagnetism using differential equations within a continuous space-time. Maxwell's electromagnetic field was defined at each (dimensionless) point in space-time. His differential equations themselves were expressed using the familiar x, y, and z coordinates of school mathematics.

But Maxwell's discovery was based on the earlier work of André Ampère, Michael Faraday, and Carl Gauss; their experimental results and theories were expressed in what turns out to be topological language. Those early researchers had not realized they were using topological terminology when they spoke of the number of "lines of force" that enter and leave an electrical circuit—a purely topological notion.

Bohm and his collaborators returned to these original insights, reformulating them with a cohomological algebra (using concepts like chains and cochains, cycles and cocycles). In this way the richness inherent in Maxwell's equations was made manifest without any recourse to underlying space. Bohm had made a first step toward a radically new language for physics.

The mathematical language Bohm adopted is essentially dual—technically speaking, the mathematical object known as a chain is complemented by its cochain. Quantum theory is also dual—observables in quantum theory are expressed in a dual way, using the wave function and its complex conjugate, or to use Dirac's technical language, a bra and a ket. The duality in both quantum theory and Bohm's algebra seemed a clue that they were heading in the right direction.

Naturally enough, applying the same approach to quantum theory was to be the next step. And once it had been reformulated, they would embark upon the unification of quantum mechanics and relativity. The research program was promising, but over the years it came up against extreme technical difficulties.

The approach of cohomology worked well in the domain of special relativity, but when they tried to extend it to general relativity and the problem of quantizing gravity, they encountered serious problems. Bohm was still trying to discover a way of resolving it at the end of his life.

It was at this point that Hiley and Bohm complemented each other. Hiley was more tolerant of mathematics than Bohm, who tended to distrust its abstract power. Their discussions were always very free and open, with Bohm constantly making interconnections and new suggestions; but often it was difficult to discern just who had suggested a particular train of thought. Generally they worked independently, coming together several times a week to discuss their progress. Following these discussions Hiley would attempt to reformulate their conclusions in a more mathematical form.

For Hiley, working with Bohm was an inspiring but sometimes frustrating experience. Meeting Bohm after many days of hard, independent work, Hiley would start to write his results on the blackboard, only to discover that his collaborator could immediately anticipate the conclusion and point out new difficulties. On other occasions, Bohm would give Hiley notes that turned out to be full of mathematical errors—yet nearly always the conclusion itself would be correct.

From time to time distinguished scientific visitors came to Birkbeck. If Bohm found it difficult to express anger within his personal relationships, he had no such problem when his passion turned to physics. On occasion, if someone disagreed with his ideas, the shouting from Bohm's office could be heard all the way down the corridor. Then one day his whole approach changed. After Bohm had given a lecture, Hiley later recalled, a physicist in the audience had kept questioning him. The two argued for a time, and the atmosphere became more belligerent until Bohm suddenly observed, "The problem is that we are not communicating. What are we going to do about it?" He then sat in complete silence. The other physicist simply did not know what to do. Bohm appears to have hoped that they could discover a nonadversarial way of working together.[2]

From that point on Bohm adopted the new tactic. In the long run, he told Hiley, arguing with people assertively is not profitable. Sometimes, to be sure, he would still try to convince others, but increasingly he realized that there was little value in persisting. Later he would say that to get his point across, he would have to talk to that person for several days, explaining his whole philosophy and metaphysics. His approach had moved so far away from the mainstream that it was increasingly rare for him to meet a physicist who was both open-minded and possessed of the philosophical and scientific background that Bohm required for a satisfying communication.

At Princeton Bohm had written *Quantum Theory;* now it was the turn for a book on special relativity. A great deal of mythology surrounds Einstein's physics, such as the remark attributed to Sir

Arthur Eddington (whom we met earlier in this book negotiating with the Devil). A reporter is supposed to have asked Eddington, "Is it true that only three people understand Einstein's theory?" to which the physicist replied, "Who is the third?"

Relativity was supposed to be counterintuitive, annihilating commonsense notions of space and time and replacing them with rods that contract and clocks that run at different rates. Bohm wanted to turn the tables and demonstrate that Einstein's was a more natural way of looking at the world than Newton's. To this end he drew on the ideas of the Swiss psychologist Jean Piaget.

Piaget spent his professional life studying how young children build up a mental picture of the world. In one experiment a toy is shown to an infant, then hidden beneath a table, only to be brought back into view again. At what age does the baby develop the concept of its continued existence? How are its notions of causality, space, and time developed? Bohm saw the relevance of Piaget's work for relativity and quantum theory—and for casting them in commonsense terms. Take the supposed nonintuitive nature of Einstein's unification of space and time into space-time. Bohm argued that our most primitive perceptions are of pure movement. Motion, as a process in space-time, is both scientifically and conceptually prior to the idea of a separate space and separate time. In his 1965 book *The Special Theory of Relativity* Bohm showed that Einstein's theory reflects a world in harmony with our earliest and deepest modes of perception. Our earliest experiences are of a unified world in which space and time are perfectly integrated and in which movement is prior to stasis. Only later, Bohm argued, do we learn to separate space and time because of our existence within a macroscopic, "classical" world.[3]

Armed with his knowledge of Piaget's approach, Bohm realized that the history of mathematics and physics is the excavation of our mental processes into our very earliest perceptions. Bohm's search for a new order in physics is itself a case in point. Rejecting Cartesian geometry in favor of topology, Bohm had moved further into algebriac topology—the set of relationships that exist prior to space. This sequence is the exact reverse of the way in which children build up their perception of the world: they first perceive pure transformational relationships (algebraic topology or cohomology), then they distinguish intersecting from nonintersecting figures. Finally they come to an intuitive understanding of geometrical figures, distinguishing squares from triangles (topology). Thus in seeking an order of prespace, Bohm was also reaching back to one of the earliest ways that our consciousness structures the world.

One can only move toward these deeper levels of cognition by overcoming layers of psychological conditioning. Krishnamurti had once told Bohm to "start with the unknown." This approach meant holding sufficient energy that old forms dissolve and the mind breaks through the boundaries of knowledge. It is exactly what Einstein had had to do before he could develop his special theory of relativity. Rather than spending all his time thinking about physics, he read the philosophers Kant and Hume so that he could free himself from the conditioning of two hundred years of Newtonian perception.

Bohm sent the manuscript of his book on relativity to the American publisher W. A. Benjamin, whose referee wrote: "How wonderful it is, in a life of many manuscripts, to receive a mature book from a fluent scholar . . . the author is a playwright as well as a scientist!" He felt Bohm's conclusion—that the enterprise of science is one of perception rather than of the accumulation of knowledge—should be placed on the desk of every teacher.[4]

The Special Theory of Relativity, which appeared in 1965, so impressed Donald Schumacher, a young American studying at the Bohr Institute in Copenhagen, that he traveled to London to become Bohm's doctoral student.[5] Schumacher had made a deep study of Niels Bohr's writings; after talking to him, Bohm realized that *Quantum Theory,* which he had believed reflected Bohr's philosophy, was in fact closer to that of Pauli.

It is a truism to point out that classical physics allows for an objective description of nature. Properties exist independently of any act of measurement or observation. In his book *Quantum Theory,* Bohm had tried to show how in every observation the system and the laboratory apparatus used to measure it are irreducibly linked by an indivisible quantum of action (that is, a quantum of interaction).

Being indivisibly linked means that the apparatus and the system cannot be distinguished from one another until the measurement process is completed and a definite result registered. If the quantum could be broken apart, analyzed, or dissected, then it would be possible to say, for example, that ninety percent of the quantum is associated with the apparatus and ten percent with the observed system. In this way physicists could calculate, and compensate for, the effects of an act of observation. But the quantum cannot be broken apart; in every measurement the system and the apparatus are an indivisible, unanalyzable whole. The only way to talk about a quantum measurement is within the context of the entire experiment.

Quantum systems do not exist in the abstract. It is not correct to talk of position or momentum as being "possessed" by a quantum particle, or to say that these properties exist independently of an observation. Physical properties must always be defined within the conditions under which they are observed. How difficult it had been for Bohr to communicate such a subtle insight! No wonder, Bohm realized as he talked with Schumacher, that Bohr's scientific papers and essays all seemed to revolve around this one essential point.

Ideas of wholeness and the maxim that "everything is connected" have become catchwords in the so-called New Age. Yet as used by Bohr, Bohm, and Schumacher, they were in fact deep and subtle notions—a far cry from the belief that "we all make our own reality." Bohm was deeply aware of the connection between the way we communicate, our physical disposition and intentions, and our perceptions of the world. In turn, he had examined the movement of thought and the way we build up notions of reality. Yet at the same time he always held that something exists independent of these mental processes—or rather, that at the deepest level matter and thought cannot be separate.

The whole question of the independent (in space and time, that is) nature of reality had been the bone of contention between Bohr and Einstein. Einstein had kept insisting on "independent elements of reality." They disagreed to the point where finally the two men could no longer communicate. At one time they had interacted so intensely that they felt love for each other. Why had these two giants of twentieth-century physics not been able to come to agreement? Bohm used to tell of their last meeting in Princeton. The university gave a party in Bohr's honor and assumed that Einstein would wish to talk with him as before. But Bohr and his associates stood at one end of the room while Einstein and his colleagues remained at the other. Why, Bohm now asked as he talked to Schumacher, had a relationship based on love and the desire to pursue truth failed so miserably?*

Theirs was not a simple disagreement of personalities. Rather, it symbolizes the present fragmentation of modern physics into two irreconcilable theories—relativity and quantum theory. Was there an even deeper reason for their breakdown in communication? Bohm and Schumacher wondered. In addition to the formal mathematical language of physics, Bohr and Einstein had used informal, everyday language to speak about their ideas. While physicists define their mathematical language with great care, ordinary spoken language is normally taken for granted. It was at this level that Einstein and Bohr parted company. Without realizing it, the two men were using words and concepts in subtly different ways. The confusion and fragmentation of physics, Bohm came to believe, had its origins deep within language itself.

Language now became his passion. He studied it with as much vigor as he had previously studied physics. Schumacher asked if any concept can be communicated in which the *form* of expression truly *matches* its content. Quantum theory denies independently existing objects and properties, which means that there can be no "it" or "the" in the quantum world. For a time, Bohm and Schumacher attempted to talk without using such parts of speech. Since even nouns themselves represent isolated objects, much of their interaction ended up in active silence.

While Hiley was party to their discussions of Bohr's philosophy, he did not join in their more metaphysical discussions of language. From time to time, however, he would walk into Bohm's office and observe what was going on. Bohm would say something, to which Schumacher would nod his head. Sometime later Schumacher would make a remark, and Bohm would nod his head as if in perfect agreement. Yet to Hiley they appeared to be contradicting each other. Finally both men reached the point where they wondered if anything

* There is a certain irony in Bohm's interest in this breakdown of communication. So many of the deepest relationships in his life—with Oppenheimer, Krishnamurti, Biederman, and Schumacher—were to end in misunderstanding and painful rupture.

significant could be said about anything. They grappled in silence; one seemed about to speak, only to pull back, realizing that no possible remark could be adequate for their intellectual impasse.

Hiley's account evokes Sherlock Holmes's account, related to his friend Dr. Watson, of a fateful meeting with his archenemy Professor Moriarty. Moriarty enters Holmes's rooms, and the great detective addresses him:

> " 'Pray take a chair. I can spare you five minutes if you have anything to say.'
>
> " 'All that I have to say has already crossed your mind,' said he.
>
> " 'Then possibly my answer has crossed yours,' I replied.
>
> " 'You stand fast?'
>
> " 'Absolutely.' "[6]

Bohm's and Schumacher's work together crystallized into a paper, "On the Failure of Communication Between Bohr and Einstein." Although Bohm referred to this paper in other publications, for some curious reason it was not published until after his death.

Such an intense interaction led to a particular closeness between the two men. Saral felt that Schumacher was almost a replica of her husband, and that just as Bohm had tried to impress his own father, Schumacher would sometimes attempt to impress Bohm.[7] For his part Bohm considered Schumacher to be the closest to genius that he had ever encountered in a student and wanted to introduce him to Krishnamurti. Bohm expected a great meeting of minds.

The encounter took place in Saanen, Switzerland. Schumacher met with Krishnamurti before the Bohms arrived, and while Schumacher was deeply impressed, the meeting did not arouse much enthusiasm on the part of the Indian teacher. As they talked, Schumacher noticed that Krishnamurti emphasized absolute words like *never, totally,* and *utterly*. It seemed to him that this sort of language was getting in the way of what he was trying to say. He even wondered why Krishnamurti himself had never questioned his particular use of language.

Despite his great respect for Krishnamurti, Schumacher pointed out these observations in a gentle way to Bohm. When the three of them met together, things did not go according to plan. Alfred Korzybski had said that "the word is not the thing." Now, as he spoke with Krishnamurti, Schumacher argued that in a certain sense the word *is* the thing. Words affect thoughts and emotions; they are real things in the world and induce changes in the operation of the brain. But as Schumacher pressed the point, Krishnamurti kept repeating "The word is not the thing," and an emotional pressure built up to such a extent that dialogue broke down.[8]

In some ways this meeting was a watershed for Bohm, for while he must have already noticed Krishnamurti's way of speaking, he had not openly acknowledged it. From now on it became another nagging question that he harbored about Krishnamurti. It also exposed a highly charged emotional rift between these two important people in Bohm's life. Whenever he recounted that meeting, it was always with a degree of emotion; at times he portrayed Donald Schumacher as overexcited or possibly even unbalanced.

Schumacher catalyzed Bohm's further explorations of the role and structure of language. Indo-European languages—like English, French, and German—rely heavily upon nouns, Bohm noted. We speak and think in terms of categories and objects in interaction. To Bohm, perception and communication were indivisible, which means that we also perceive a world composed of localized objects in interaction. This idea is clearly at odds with quantum theory, which speaks of process and transformation rather than object and interaction.

Our earliest perceptions of the world are of transformation and flow. Clearly something happens to us by the time we reach adulthood. In Bohm's opinion, the culprit is language. Western society thinks, perceives, and communicates in a way that does not cohere with the actuality of the world.

Ideas on the interconnection of language and worldview had earlier been discussed by the American linguist Benjamin Lee Whorf (in the Sapir-Whorf hypothesis) and the Russian psychologist L. S. Vygotski. Bohm did not appear to know of this work, and his own researches led him to develop a hypothetical verb-based language he called the rheomode. If only we could communicate using the rheomode, rather than in English, German, or French, then our perceptions and the form of our communications would cohere more closely with their content.

Bohm discussed these ideas at Brockwood Park, gave two talks at the Institute of Contemporary Arts, and circulated a draft paper among friends and colleagues—which later was to become a chapter in *Wholeness and the Implicate Order*.* For a time a small group at Brockwood adopted the rheomode as an experiment. After one of their rheomode-speaking sessions, Bohm noticed that everyone felt as if they were on one side of a gap, trying to reach the other. It would be helpful, he suggested, if they said, "There is no gap, we are on the other side." For several weeks some members of the staff took this literally, believing they were truly "on the other side"—a form of one-upmanship that caused annoyance to other members of the school.[9] The more perceptive noticed how, over time, the group began to use the rheomode's verbs as stand-ins for nouns, defeating the very purpose for which the language had been created. Transforming language was a project more difficult than Bohm had anticipated.

Bohm's ideas on the rheomode were intriguing, yet the response he got from most professional linguists (but not all) was discouraging; they found his arguments naive and unconvincing. By way of qualification it should be added that during this same period the work of Noam Chomsky and his followers was in fashion, and ideas like the Sapir-Whorf hypothesis were largely ignored. After presenting his ideas in talks and a publication, Bohm no longer pursued the rheomode.

In the last year of his life, however, Bohm met with a group of Native Americans who were all speakers of the strongly verb-based Algonquian family of languages. (The participants included Blackfoot, MicMaq, Cheyenne, Ojibwa, and Soto.) Bohm was struck by their process-based vision of the world and by the way they themselves viewed the role played by their language. Here was a society, it seemed to him, that practiced what he had envisioned for his

* It was Bohm's custom to circulate his more philosophical papers in draft form for several years, receiving feedback and constantly making modifications before final publication.

rheomode. Two linguists, Alan Ford of the Université de Montréal, and Dan Alford of the California Institute of Integral Studies, both of whom studied Algonquian languages, were present at the meeting and were sympathetic to Bohm's arguments about the rheomode.[10]

Schumacher, for his part, was dubious about the importance of the rheomode and tried to explain to Bohm the position taken by Ludwig Wittgenstein in *Philosophical Investigations*. To Schumacher's surprise, Bohm refused to read Wittgenstein. Just as communication had broken down between Bohr and Einstein, difficulties were now developing between Bohm and Schumacher.[11]

Although Schumacher pursued ideas with great involvement and excitement, as time went on his intensity compromised his mental equilibrium and led to a series of breakdowns. As Bohm began to notice signs of Schumacher's distraction, he became extremely troubled and entered into an agitated state himself. As Saral put it, "He was terribly upset by Donald. He didn't know what to do, a good mind going to pieces."[12]

It is difficult to present an unbiased account of what happened next since emotions were so heightened. Deeply upset at Schumacher's condition, Bohm would tell Saral about his distress at the loss of such an exceptional mind, of someone with whom he had done highly creative work. The Bohms certainly visited Schumacher in the hospital and spoke to the young physicist's father. On the other hand, the proximity of this mental breakdown may have stirred up in Bohm painful memories of his mother. He may have wondered about the ultimate prognosis of his own periodic bouts of depression. In any event, Bohm began to distance himself from the younger man. Admittedly, during his breakdown Schumacher exhibited symptoms that caused considerable worry to his friends and colleagues. Nevertheless, Schumacher himself was shocked at what he described as a "vigorous repulsion" on Bohm's part. He felt stigmatized by an attitude that was "like something out of the Middle Ages."[13] Even Hiley was struck by the abrupt way Bohm cut himself off from his former friend and colleague.[14] Before leaving for the United States, Schumacher made a final visit to Bohm's office. The interview was particularly painful; Bohm appeared full of recriminations toward him.

Mental instability has a tendency to spill over into the environment, and Bohm may simply have been attempting to distance himself from its infectious quality, albeit in a somewhat insensitive way. Yet decades later Bohm's memories of Schumacher's breakdown were still highly charged, and I could not help feeling that he interpreted it as yet another betrayal.*

With Schumacher gone, Bohm continued to work on physics with Basil Hiley and on the weekends drove down to Brockwood Park. One day a letter arrived from an old friend who opened up the question of his return to the United States. Rossi Lomanitz, who for many years had worked at laboring jobs, now had a position at the New Mexico Institute of Mining and Technology. Would Bohm like to join him there?

Bohm reacted with caution. He was content where he was and did not believe that the American political climate had changed in any fundamental way. Nevertheless, Stephen A. Colgate, president of the institute, wrote to Spurgeon M. Keeny, Jr., of the Presidential Scientific Advisory Committee, "to encourage the government to offer the privilege to David J Bohm of return of his passport and entry into this country."[15] Bohm had refused to testify before the HCUA, he said, in order to protect friends "from guilt by implication." Since that time, others who had done the same had returned to useful lives as scientists in the United States, but Bohm was still exiled from his native land.

* The writings of the psychiatrist R. D. Laing were much in fashion in the late 1960s and early 1970s. Laing developed a radical form of treatment whereby psychotics were allowed to journey into their innerworlds without restraint or medication. Some of Laing's theories were based on Gregory Bateson's theory of the double bind. Not yet knowing the facts about Bohm's family background, I presented Bateson and Laing's position to him, using a particularly unfortunate example.

Suppose, I said, each time a young child approaches his mother, he is frightened or repulsed by her reaction. The child's natural reaction is to flee, but as soon as he does so, he feels insecure and desires to return to his mother—once more to face her irrational reaction. The result is an unbearable and irreconcilable double bind that, in a sensitive individual, can lead to schizophrenia. Bohm's reaction shocked me by its uncharacteristic harshness. "Laing is far too easy on those people," he said. "The whole thing is a failure of intelligence. The person does not see that these conditions no longer apply." His tone was so irritated that it left no room for debate. Bohm then referred to Donald Schumacher's condition as a failure of intelligence—as if Schumacher's illness, indeed all mental illness, were a moral failing on the part of the sufferer.

Keeny's reply was not encouraging. It was not up to the government to make an offer to Bohm, he said. "If Dr. Bohm is interested in returning to this country, I would suggest that he obtain the services of a good lawyer and present his petition in writing."[16]

Colgate persisted. He wrote to the assistant chief in the legal department of the State Department's Passport Office, to the director of the Passport Office, and to the deputy director of the Visa Office. He also requested legal advice from the Federal Bar Association. What exactly was Bohm's legal position? he asked.

Bohm's passport had been removed in Brazil on the grounds that "current passport regulations did not permit use of passports by persons charged with or suspected of communist membership or affiliation except for travel to U.S." But it was not clear that the passport had been taken in the proper manner. Bohm had not filed a certificate of renunciation of his citizenship, and the U.S. consul in Brazil still accepted him as an American citizen until they learned he had been naturalized as a Brazilian on November 23, 1954. Colgate warned Bohm not to discuss the matter with the U.S. consul general in London until a full legal assessment of his position had been made.

In the end Bohm decided not to pursue the matter. He wrote to Lomanitz that the American authorities would look favorably upon him only if he published an article of an "anti-communist" nature. To Bohm, this was "not really compatible with dignity."[17] He could say that "Communism is no solution to the problems of mankind, but . . . nevertheless, the problems that originally made me look to communism for a solution are even more pressing than they were 25 years ago." Once he felt that Communism "might" be the answer, but now he saw that "communism is not the answer, and that in its denial of freedom, it is in a crucial way, worse than the evils it aims to combat." Yet "I feel it wrong to say it in order to regain US citizenship. For then, I am saying something not mainly because I think it is true, but rather, for some ulterior purpose." It would be like writing a scientific paper in order to impress one's superiors or get a better job.

In view of this lofty disregard for his own interests, it comes as something of a disillusionment to learn that several years earlier, on March 23, 1960, Bohm had already made such a statement for the American authorities. The document is an apologia for his earlier beliefs, explaining how he had been a member of the Communist party for nine months, thinking that it offered a solution to the social and political problems of the time.

"I was also passing through a difficult period in my personal relations. I felt lonely and isolated. In addition I had lost much of the feeling of purpose in life that I had had before. Just at this time I came into contact with some Communists. . . . these people had a unity and comradeship based on a common purpose of creating a better society. . . . Of course, I still had considerable misgivings about the dictatorial methods of the Communists but their new policy of abandoning the principle of violent overthrow of existing governments helped to reassure me in this regard."

As to his support of Russian Communism: "It must be remembered that at the time the Russians were fighting very effectively on our side against the Nazis." Yet as the war came to the end, his statement declares, he realized that "the basic idea of Communism is false." His statement goes on to cite the loss of freedom, the brutality, and the torture associated with the Communist regime. "The furthering of Communist power is fundamentally wrong, because it is bound to lead . . . to the corruption of all principles." At times Bohm's style is not unlike that of his former nemesis, President Dodds of Princeton University!

His 1960 statement raises puzzling questions. Why did he place his rejection of Communism at the end of the Second World War when in fact his letters from Brazil are staunchly pro-Communist? And why, even into the 1970s, did he continue to tell his friends that, even if it were the only way to regain his American citizenship, he would never write the sort of statement they wanted?

CHAPTER 14
The Implicate Order

AT THE END of the 1960s, the talk in Bohm's group at Birkbeck had been about topology, nonlocality, prespace, and a new order in physics based on similar differences and different similarities. Then suddenly—for Hiley at least—a new idea emerged from Bohm, apparently entirely out of the blue. Bohm called it "the implicate order." Years of thinking about canonical transformations, the infinite nature of the cosmos, and the need for a new order in physics had finally coalesced into a vision of a hidden enfolded order. All that we see around us, Bohm declared, the province of Newtonian physics, is nothing but the world's surface, its explicate order. There lies something far deeper, out of which our explicate world unfolds. Bohm called it the implicate order, an order of process and transformation. Over the next years, in lectures, essays, and a book, *Wholeness and the Implicate Order,*[1] Bohm expanded this idea beyond the immediate domain of physics, making it available to a much wider audience.

It was during one of his most productive periods, the summer of 1971, that I first met David Bohm. One day—the sixteenth of June, to be precise—while I was on sabbatical with Roger Penrose, I looked in on the physics graduate room at Birkbeck College. An active discussion was in progress, with a young Canadian student claiming that everything was relative, while an older man in a tweed jacket, in true Socratic manner, pushed him toward paradox and revelation. "You say there is no absolute. Is that an absolute statement?" he asked.

The older man was David Bohm, who had just returned to

London after attending Krishnamurti's talks in Saanen. Never before had I encountered such passion and clarity. Here was a mind of great power housed within a frail body. Apart from a fluttering of the hands as he spoke, it was as if all his energy were reserved for the mental task of pursuing truth, no matter where it took him.

When Bohm left the room, I followed him down the corridor and told him that I wanted to talk to him. I must have said this with a great deal of urgency, for he immediately invited me into his office. From that day onward we met regularly. The arrangements for our meetings were made in an indirect "English" fashion. I would telephone Bohm at his home and allude to a particular topic. In reply, he would glance toward the fact that he would be in his office the following afternoon. Several times a week, over the next eight months, we talked through the afternoon and on into the early evening, when I would walk Bohm to the Goodge Street underground station for his train to Edgware.*

While some people found Bohm reserved and difficult to reach, I felt an immediate rapport with him. Over these months he guided me deeper and deeper into his thinking, not revealing everything at once but waiting for me to invite him to take the next step. Our discussions began with quantum physics and ranged through causality, time, mathematics, and order. One night, while I was thinking over all he had said, it struck me that the essential, unstated point in our discussions was the nature of the thinker and the thought. This, I sensed, was the true essence of his philosophy, and excitedly I telephoned him to say, "We've got to talk about mind."

Little did I know that this was exactly what Bohm had been investigating over the years. Now our discussions went into the field of consciousness and language. Yet at no time did Bohm mention the name of Krishnamurti. It was only from a postdoctoral student, David Schrum, that I learned of Bohm's connection with the Indian teacher.

In some ways this reserve was characteristic of Bohm, who complained about fragmentation both in science and in life and yet, with respect to his friends and colleagues, compartmentalized his areas of interest, discussing physics with one, Krishnamurti with someone else, and with yet another, language and perception. After his death a

* I kept a notebook in which I recorded several of our discussions and conversations.

number of people who had known him quite well during his period in England were surprised to learn that he had ever been a Marxist.

Before I came to Birkbeck, I too had been interested in canonical transformations and was now excited to learn about Bohm's implicate order. For the boy crossing the stream in the woods, reality had been transformed into an endless flowing movement. At Berkeley he had imagined the electron as a continuous process of collapse and expansion. In Brazil nature had been an endless series of levels. In Israel and Bristol movement and transformation were powered by the dialectic, and in the field of Newtonian physics, this manifested itself as canonical transformations. Finally, at Birkbeck, it became the implicate order.

The cosmos is inexhaustible—science only touches its surface. Yet because of the harmony among its levels, we make sense of the world. Before the advent of quantum theory, science dealt with only one level. Bohm called this level the explicate order, the familiar world of independent, well-defined objects. The explicate order is the order of space and time, separation and distance, mechanical force and effective cause.

But there is a deeper order, one closer to our unconditioned thought and perceptions and more in keeping with the implications of quantum theory. Bohm called this the implicate, or enfolded, order. The implicate order lies beneath the explicate and gives rise to it. Yet in another sense all levels are aspects of a universal movement that Bohm called the holomovement.

Despite the radical scientific revolutions of the twentieth century, physicists still tend to think in terms of the explicate order. They dissect matter, looking for smaller and smaller entities and an ultimate elementary particle. Their fundamental equations are written using the coordinates of space and time. And their world of physics is distinct from that of mind. By contrast, Bohm believed that the unity of matter and consciousness is radically different; it is the world of transformation and flowing movement described within the implicate order.

Bohm sought a holistic physics. Indeed, the ultimate ground of the implicate and explicate orders, the holomovement, is the movement of the whole. What is the nature of this whole, this totality of matter and consciousness? To understand the answer, it is necessary to know what Bohm meant by "reality."

That word *reality* is the pawn in a wide variety of language games. The naïve scientific realist uses it in the sense of things being "real"—that is, objective, independent, autonomous, and located in space and time. Idealists believe, by contrast, that reality has its origins in universal mind. Then there are those who, with a superficial knowledge of psychology and a misunderstanding of quantum theory, believe that "we all create our own reality." Bohm's own approach transcended this philosophical duality, making a fundamental distinction between "reality" and what he termed "that which is" or "all that is."

The cosmos is fundamentally independent of our wishes, thoughts, and desires, he believed. In this sense, the sense he shared with Einstein, it is objective. Opposing the Copenhagen interpretation, Bohm held that an observer-free account of a quantum measurement is possible. This objective nature of the world Bohm referred to as "that which is."

"That which is" includes not only the material universe but consciousness, the human observer, and many other levels and processes of which we are at present ignorant. The question then arises as to how we come to know anything about "that which is." Our perceptions and transactions with the world, as well as the mental maps we create, are all influenced by the processes of the mind and its physical structure. Our perceptions are conditioned at the individual, social, and linguistic levels. It is this shared perception, this knowledge of "that which is," that Bohm referred to as "reality." "That which is" is inexhaustible and infinite in its potential; "reality" is only an aspect of this much wider domain.

"Reality" has a subjective element, yet it is not exclusively subjective. We are free to believe anything we wish about the reality, but if we push a belief too far, we discover areas of our socially shared theories that do not cohere with experience. Like Dr. Johnson kicking the stone to refute Bishop Berkeley, we come up against the grain of the world and are forced to work with and not against it. "Reality" therefore has both objective and subjective elements.

The investigation of "reality" is the province of science. For four hundred years reality had been described in terms of Newtonian physics, as a collection of separate interacting objects moving in space and time. This explicate order, however, subsists on an underlying implicate order.

What is the implicate order? The notion is so subtle that Bohm chose to approach it through allusion and metaphor. He likened the explicate order to a photograph of reality. In a normal photograph there is a clear correspondence between the object and its image. Features on the original object are reproduced in corresponding regions of the photograph—light comes from each part of the object and is focused onto the photographic paper. Mathematically speaking, there is a one-to-one correspondence between object and image.

But there is another method of photography called holography, or lensless photography, in which light from each part of the object falls onto the entire photographic plate. In a holograph each part of the plate contains information about the entire scene; light in a sense becomes enfolded across the holograph.

Bohm compared the implicate order to the process of the holograph. Even when a tiny portion of the holographic plate is broken off, he pointed out, it contains information about the entire scene.* Likewise, in the implicate order the entire universe is enfolded into each part. This means that our everyday notions of space, including our ideas of distance and separation, apply only at the surface of things, within the explicate order. In normal photography the regions of the original image may be far apart while they are superimposed on a holograph. What if the cosmos itself is closer to a holograph than a photograph? What appear to be distant objects will, within the implicate order, lie close together, the one being enfolded within the other.

Two knots are tied far apart on a piece of string. If the string is folded into a bundle, the two knots touch. Points that are far apart in a linear, explicate order can be close together in an enfolded order. Moreover, within the implicate order A may be enfolded within B, at the same time as B is enfolded within A. In this sense objects that, within the explicate order, are far apart may, within the implicate order, be mutually enfolded. There are many suggestive connections

* "Holographs" sold in stores today use a variety of techniques that are not always like those of the original holographs Bohm used in his example.

between this notion of an implicate order and Bohm's other work on topology and prespace.

Bohm took pains to point out that his holographic example was but a shadow of what he meant by the implicate order; nevertheless, the idea entered the popular imagination in the sense of "the universe is a holograph." The neuroscientist Karl Pribram was struck by the appropriateness of Bohm's metaphor as a model for memory and the brain's function. An outstanding problem in neuroscience is "the search for the engram," or the way the brain encodes memories. The brain does not operate like a card index system or a computer memory. When it is damaged—either in a specific area or peppered across larger regions—specific memories are not lost, the way cards are removed from a card file or pages from a book. Moreover, skills, which would be stored in one region of the brain, are readily transferable to other regions. In a right-handed person the task of signing one's name is stored in the left hemisphere of the brain. Nevertheless, most right-handed people can perform the bizarre party trick of writing with a pencil held in the toes of the left foot. How is such a specific motor skill transferred from one side of the brain to the other?

Pribram was attracted by Bohm's observation that even a tiny portion of a holograph contains information about the entire scene. He proposed a "holographic model" of the brain, along with a theoretical explanation of how incoming neural impulses are enfolded and unfolded across the brain's surface.[2]

Another of Bohm's metaphors sprang from an experiment he had once seen on television. In the experiment glycerine is held between two cylinders, the inner of which is free to rotate. A drop of ink is placed in the glycerine, and the inner cylinder is slowly rotated. As the cylinder turns, dragging the glycerine with it, the drop elongates into a thread, which is drawn out until it becomes so fine that it is no longer visible. But when the rotation of the inner cylinder is reversed the thread reappears and collapses back into the original drop.

Bohm pictured the ink drop as the explicate order and the invisible, enfolded thread as the implicate order. The image could be taken even further. Suppose, he said, the cylinder has turned n times. Then another drop is placed close to where the first was added. After an additional n turns, a third drop is added, and so on. Upon reversing the cylinders, a drop appears out of nowhere and seems to move—as a series of droplets—along a track between the cylinders. The whole process has the appearance of an elementary particle moving within a bubble chamber. This was close to Bohm's original image of an elementary particle as a wave that collapses inward from the entire universe, expands outward, and collapses in again. It suggests that what is perceived at the explicate level as an elementary particle is, at the implicate level, the manifestation of a constant process of enfoldment and unfoldment.

But again the ink drop analogy is too simplistic and mechanistic. Its major drawback is that it implies that the implicate and explicate orders are, like canonical transformations, simply transformations one into the other. What Bohm had in mind was that the implicate is somehow "deeper" and more fundamental than the explicate. While the explicate unfolds from the implicate, then folds back again, it is impossible to unfold the totality of the implicate into the explicate. Only a part of the implicate can ever be unfolded at any one time; the rest remains inaccessible to our explicate world. This evokes Bohr's notion of complementarity, whereby only the wave or the particle nature of the electron can be manifest at any one time. The underlying implicate order of the electron is unknown; all we see is its wave, or particle, appearance in the explicate order of laboratory experiments.

The connection between this idea and Bohr's complementarity is a subtle one. Bohr's argument was about the limitations of *descriptions* of the subatomic world. Bohr generally avoided any discussion about the ontology or underlying actuality of quantum processes. On those rare occasions when he did venture an opinion, it was subtle to the point of obscurity. By contrast, Bohm's idea, that at any one time only a limited aspect of the implicate order can be made explicate, becomes almost an ontological statement about reality.

Other connections suggest themselves, particularly the relationship of the implicate order to the collective unconscious of Carl Jung, rather than to the unconscious as described by Freud. For Jung the unconscious has a rich and deep structure, much of which can never be made manifest in conscious awareness. As an example, the archetypes, which Jung believed give structure not only to individual behavior but to that of a group or nation, can never be directly apprehended but are presented to us only in their particular manifestations as symbols and structures in art, dreams, and culture.

In discussing the implicate order, Bohm was forced to work through metaphor—holographs and ink drops. This is a useful way to proceed when attempting to explain new and difficult ideas. But danger lies in the fact that the metaphor is both similar to and different from that for which it stands. Using an image as a stepping-stone does not normally pose a problem, since one can then go to deeper levels of explanation and approach the concept more precisely. In physics metaphors are eventually replaced by equations and experiments. The difficulty Bohm faced was: What exactly lies beyond the image? When speaking to a scientific audience, he would refer to the Green's function, a mathematical device in which solutions are enfolded together. But again the Green's function is not the implicate order, it merely points toward it. It is yet another metaphor.

A further difficulty arises because, in terms of its intellectual origins, the implicate order is a river fed by two sources—Bohm's vision of an infinity of levels in the cosmos and his search for a new order in physics. In the latter stream, the implicate order is a descriptive device or image; in the former it says something about the way the universe is actually constructed. In his lectures, writings, and informal conversations, Bohm moved with ease between these two uses. As describing a new order, the implicate order showed how physics is still caught up in a restrictive conceptual and mathematical language for discussing the universe. What we take as space and particles, Bohm argued, are simply modes of description that focus on particular aspects and do not exploit the richer ground of experience made manifest through quantum theory.

In this sense the implicate order is a new way of seeing and talking about the world. It directs our attention away from boundaries and independent existences into holism, interconnectedness, and transformation. It argues that explicate order descriptions can never exhaust physical reality. The implicate order is a door into new ways of thinking and the eventual discovery of new and more appropriate mathematical orders. It is both a philosophical attitude and a method of inquiry.

But as Bohm sometimes used it, the implicate order can also refer to a particular level of reality, a subtle material realm beyond the domain of particles and forces. In this sense order is identified with essence. Just as chairs and tables reveal an underlying atomic existence, so too will space, time, and elementary particles eventually reveal an underlying reality of enfoldment. The implicate order is now a hypothesis about the nature of the physical world, a suggestion that may one day be put to experimental test. In this sense it is even possible to speak of "the" implicate order, as if it were some particular level of existence not appreciable by our ordinary senses. When we were stirring sugar into our tea at Birkbeck College, we could joke, "Ah, the sugar's entering the implicate order."

When *Wholeness and the Implicate Order* appeared in 1980, it had an immediate appeal—not so much to physicists and philosophers, who were somewhat suspicious of its ideas, but to writers, artists, musicians, psychologists, and others who felt they had always experienced the world in this way and now had access to a powerful common metaphor.

The book excited the painter David Hockney, who described himself as "ready for Bohm's ideas." Hockney at that time was moving beyond the confines of traditional perspective and attempting to introduce time into his paintings, together with a new experience of space. Bohm, he realized, was speaking about the interconnectedness of things and about the way consciousness enters into our notions of reality. As Hockney read the book, he found that he could translate Bohm's ideas directly into his own visual language and perceptive experiences. Some years later Hockney met the Bohms during a lecture the physicist was giving in California. Unfortunately at that time Hockney had not yet been fitted with new and more powerful hearing aids, which made communication between the two men disappointing.[3]

The writer Ian McEwan acknowledged *Wholeness and the Implicate Order* in his novel, *The Child in Time*.[4] McEwan heard Bohm speak at St. John's Church, Smith Square, in London and was struck by his strong moral and social sense and by his degree of spiritual insight. As to *Wholeness and the Implicate Order*, McEwan found the science forbidding and managed to read only about one quarter of it. But he did mine it for ideas, metaphors, and suggestions, as well as for the preoccupation of his character Thelma.[5]

A central event in McEwan's novel, the kidnapping of a three-year-old girl from her father at a supermarket, resonates through the book so as to link events from past, present, and future—events, moreover, that do not lie within each other's causal domains. This structure is very much in the spirit of Bohm's implicate order. Yet McEwan himself says that he did not pick up the particular approach from Bohm. Hidden patterns and structures, he believed, have always been the province of the writer. While Bohm's approach was congenial, it was, to McEwan, hardly novel.[6]

To many others, the implicate order, as a striking new image of the world, became a powerful talisman that enabled them to enter new areas of thought and knowledge and express them in new terms. It did not really matter if the popular definitions were neither philosophically rigorous nor mathematically precise; the important thing was the sense of liberation inherent in abandoning the old explicate order for the creative possibilities of the implicate order. Outside technical physics the implicate order became Bohm's best-known idea.

The early 1970s also saw a revival of Bohm's twenty-year-old hidden variable theory. He himself had long since lost interest in his causal interpretation, in favor of new orders in physics. It was left to Chris Philippidis, Bohm's graduate student, to revive the idea. Philippidis, who had left-wing interests, enjoyed discussing Lenin's writings on science. He also discovered Bohm's interest in Wilhelm Reich, the psychiatrist who had developed theories of physical armoring of the body, the importance of the orgasm, and the connection between the rise of fascism and sexual repression. Hidden variables were never part of these conversations.[7]

At the time Philippidis was supposed to be working on the ESAB effect, discovered in Bristol by Bohm and Aharanov, but he was blocked in his research. Reading over the recently published correspondence between Einstein and Max Born, he came upon a discussion of Bohm's hidden variable theory. Then, while at a conference in Italy, he learned to his surprise that several continental physicists took this work seriously. As a result, and despite being cautioned by friends that the whole idea was rubbish, he decided to study Bohm's 1952 papers on hidden variables. The next step was to confront Bohm himself, who was more interested in talking about algebras but, when pressed, admitted that he had made a tactical error in his original presentation of the theory. The term *hidden variables,* he said, created the wrong impression, and the papers themselves were too rigid and deterministic. He had badly misinterpreted the lack of reaction from the physics community, not realizing the social and economic pressures under which most physicists work.

In the early 1970s Bohm had little interest in reviving his theory, and apart from discussions with Hiley, Philippidis essentially worked on his own. Electrons move along paths that are determined by the quantum potential, the younger man saw. The potential itself is enormously complicated, for it contains information about the entire experimental organization. Philippidis attempted to work out the shape of this quantum potential, producing page after page of calculations. A friend, Bob Kay, who worked at Guy's Hospital and medical school suggested that the whole thing could be solved by computer. A great deal of computing time was needed, so Kay and Philippidis sat in a pub each evening until around ten, when they could work on the computer overnight.

As the first results appeared, Philippidis saw that the quantum potential was like a landscape, with the electrons running down valleys or channels. He showed the results to Bohm who, after taking only one glance, told him that a minus sign was missing. Instead of running along channel bottoms, the electrons should move along their tops. Philippidis quickly discovered the error, and with the help of a computer package that was used by geologists to produce landscapes, he plotted out the entire quantum potential.

Philippidis discussed his results with Basil Hiley and Hiley's student, Christopher Dewdney. Giving Dewdney a copy of the computer program, he and Kay suggested that Dewdney should calculate the trajectories that went along with the quantum potential.

As Philippidis and his colleagues produced their computer plots, Bohm became more interested in reviving his own earlier theory. Hiley, interested in how the quantum potential changes over time, had Dewdney calculate what happens when an electron approaches a barrier. Dewdney also make a short sixteen-millimeter film showing the movement of a "wave packet" of electrons.

Philippidis now applied the approach to his original dissertation problem, the ESAB effect. Within the conventional interpretation of quantum theory, the ESAB effect is nonlocal, since an electron's interference pattern is changed even when it is shielded from an electrical field. Within Bohm's causal interpretation, apparent nonlocality can be generated in what could be thought of as a local way because the quantum potential, which is modified by the presence of an electrical field, is not diminished by distance. Rather than demonstrating strict nonlocality, the ESAB effect is really about the holistic or global nature of quantum phenomena.

Philippidis wrote up the paper, putting Bohm's name first and adding that of his friend, Bob Kay. A few days later Bohm's corrections arrived, with the order of names on the paper changed. Philippidis was surprised since the original idea had been Aharanov and Bohm's, but with characteristic generosity Bohm pointed out that the idea of using the quantum potential in this work had come from Philippidis, and he should therefore be given full recognition.

The quantum potential also made sense out of the famous double slit experiment. In this experiment a beam of electrons is directed toward a barrier containing two slits. Because of the electron's wavelike nature, the beam exhibits the familiar interference pattern characteristic of light. And as expected, if one of the slits is blocked, no interference pattern is produced. What is puzzling, however, is that this interference pattern persists even when the beam consists of a single electron. How is this possible? How can an electron split itself in two and then interfere with itself? Common sense tells us that a single electron must go through only one slit. But how does the electron "know" when the second slit is blocked?

Bohr would have objected to loose talk about electrons "knowing" or even having well-defined paths. Yet without such informal discussion, it is difficult even to think about the meaning of an experiment. Within the conventional interpretation something nonlocal is clearly going on, but what? For Bohm the explanation was perfectly obvious. A complex quantum potential is produced by the two slits. Individual electrons travel in this potential along well-defined paths through either of the two slits. Depending upon slight variations in the initial position and speed of the electron as it approaches the barrier, it will take a particular deterministic path. No reference is therefore made to nonlocality; rather the essential wholeness of quantum phenomena is made explicit.

When Bohm and Hiley saw these computer print-outs for the first time, they found them so beautiful that they believed that physicists would at last take the causal interpretation seriously. Even if it was not the last word on the quantum world, at least it provided an intuitive picture of what was going on.[8]

Bohm made one last attempt to present his theory to the physics community. This time he adopted a more politic approach; instead of claiming that he was about to replace conventional quantum theory, he simply showed how the quantum potential could be used to explain such puzzling phenomena as nonlocality, quantum tunneling, the double slit experiment, and so on.

Bohm also decided to change the name of his theory. *Hidden variables* had clearly been a mistake for, as Bohm and Hiley later wrote, "our variables are not actually hidden"—the electron is treated as an actual material particle with a well-defined position and momentum.[9] The *causal interpretation* was a more appropriate name, but later even this one sounded overly restrictive. Only in the mid-1980s did Bohm and Hiley agree that what they were dealing with was an ontological interpretation of quantum events. Ontologies are accounts of the way things actually are. Bohr's Copenhagen interpretation offered no such account; indeed, at times it even seemed to deny the underlying reality of quantum processes. Bohm's new theory was an account of the actuality of the quantum world—hence the new name, ontological interpretation. Bohm also liked the neutrality of the word *ontological,* for it had no previous associations within the physics community.

Bohm's revived interest in his own theory had been stimulated by the computer images of particle tracks and the quantum potential. Now he encouraged his group to look at such puzzling problems as the quantum Zeno paradox, the quantum measurement problem, and a "delayed choice" experiment proposed by the physicist John Wheeler. Thanks to his ontological interpretation, Bohm and his colleagues were able to give rational explications for problems that had earlier appeared paradoxical.

In his letters from Brazil, Bohm had hoped to modify his original 1952 paper so that its predictions would differ from those of conventional quantum theory. Experiments performed over very small distances or short time intervals, he speculated, might decide in favor of his own approach. But now he discouraged such speculation, stressing that his theory reproduced exactly all the predictions of conventional quantum theory. Bohm's essential point was that, contrary to what Bohr and his colleagues had claimed, a complete and consistent alternative to the conventional theory was possible. Once this door had been opened, then other theories could follow.

Bohm's ontological interpretation had the advantage of allowing for the pictorial representation of quantum events. It was an easier way to understand and teach quantum theory, Bohm felt. Yet rather than replacing Bohr's approach, Bohm always presented both theories to students side by side. By holding a tension between them, he believed that something new, beyond the limits of both approaches, would be generated.

Basil Hiley's own interest had always been in the idea of a new order for physics, with the emphasis on algebras, topology, and notions of prespace. He had never really bothered with the causal interpretation. But now, as he discovered the tremendous interest that a younger generation of physicists were expressing, he was persuaded to devote his energies to this line of research.

If Bohm and Hiley were increasingly working on the ontological interpretation, they did not entirely abandon the search for new orders. They had already derived the equations for electrodynamics from cohomological relations. Their next step was to extend this approach to quantum theory. In particular, they became interested in the idea of prespace—an order that exists below the level of elementary particles and precedes notions of space and time. The idea was to discover sets of mathematical relationships that, at the quantum level, would reduce to what appears to be elementary particles moving in space. At the underlying prespace level, there would be no distinction between matter, space, and time.

Even as genuine insights developed, Bohm and Hiley's approach was bedeviled by technical problems. They worried that time was not present in their theory in any truly deep way. In physics time is often used as a parameter, rather than playing a fundamental dynamical role. The theoretical chemist and physicist Ilya Prigogine argued that physics dealt only with "being" and had yet to come to terms with "becoming." If Bohm and Hiley were to create a new order for physics, it had to be based within a dynamical picture of time.

In the 1970s the South African physicist Fabio Fescura joined Bohm's group as a research student. He was particularly interested in the works of the nineteenth-century mathematicians William Kingdon Clifford, Hermann Grassman, and W. R. Hamilton. As a result, Hiley came to read their original papers. All had been preoccupied with ways of representing movement, transformation, and change.

Later mathematicians had extracted only the static parts of these algebras, which they considered important, leaving behind what Hiley now discovered was pure gold. The original mathematics was very important, he realized. When he explained the ideas to Bohm, Bohm immediately realized that they had captured something of the essence of movement in a way that resonated with his own notion of the implicate order.

It had been Grassman's original goal to create an algebra for the movement of thought. Thoughts do not follow one another like a row of railway wagons, Grassman argued; rather, each thought enfolds others. A present thought enfolds a future thought and, in turn, contains traces of past thoughts. Grassman was trying to capture an algebra of pure becoming. If becoming is treated in an analytic way, as a process broken down into a succession of various components, then it no longer expresses the whole, indivisible quality of movement. For Grassman, by contrast, an algebra of becoming had to be an algebra of wholeness.

Bohm realized that Grassman's algebra of thought was pointing toward an algebra of prespace. It could even be used to derive the properties of spatial directions. He found it remarkable that an algebra originally created to describe thought should serve so well to derive the properties of space. For Bohm, it was a further demonstration of the correctness of his belief in the underlying unity of nature and consciousness.

This work on the algebras of prespace was difficult and highly speculative. As Bohm was well aware, most physicists were not interested in what he was doing—his notions of nonlocality and prespace were far too radical. In addition, the highly creative work demanded great energy. It could not be sustained on a day-to-day basis, and Bohm benefited from the stimulation he gained while traveling, giving talks, and meeting other thinkers.

Through approaches like the implicate order and the algebras of prespace, Bohm's interests in consciousness and physics were coming together. Deploring the pervasive fragmentation that thought created in the world, his vision had always been one of wholeness. A particularly pernicious fragmentation was the division between mind and matter. Over the years Bohm further developed his earlier ideas on "egoic process" and the indivisibility of matter and mind.

This interest brought him into contact with the New Age movement, particularly in the United States. In Brazil he had been strongly antipathetical toward any whiff of "mysticism," but now he began to flirt with what could generally be called the paranormal.

On one occasion Bohm received a visit from the physicist Brendan O'Regan, who was associated with the Society for Psychical Research and later became research director at the Institute for Noetic Sciences in the United States. Claims were being made, O'Regan explained, that metal (keys and spoons) could be bent using the power of the mind. Bohm asked Hiley to join in the discussion, and both of them took the proposal seriously. After all, Bohm had long emphasized the unity of mind and matter; was it possible that direct effects could be discovered? Hiley suggested that an objective test was needed, one that could distinguish psychically deformed metal from that which had been subject to normal mechanical bending. He and Bohm therefore consulted an experimental physicist in the department, John Hasted.[10]

At this point the Israeli stage magician, Uri Geller, arrived on the scene and put on a display of alleged psychic metal-bending in Hasted's office. Hiley cautioned Bohm that a theoretical physicist is not the best person to distinguish between a magician's sleight-of-hand and a genuine new phenomenon, but Bohm became excited at Geller's powers nonetheless. Joe Zorskie, a physics teacher at Brockwood Park, joked about Bohm arriving at the school bearing a key that Geller had bent, almost as if it were a holy relic. Zorskie was taken aback by Bohm's enthusiasm, particularly when, unable to find the key, Bohm declared, "The key is gone," as if this too were a psychic phenomenon. An hour later, when it was found, Bohm seemed to take this as additional evidence of the paranormal.[11]

At Birkbeck Hasted and Bohm set up experiments to test the extent of Geller's powers. The encounter left much to be desired. Geller, to be sure, could certainly diminish the masses of test substances sealed in glass vials and influence scientific measuring apparatuses. But afterward, when the two scientists examined the apparatus, they discovered that the vials were cracked and the meters incorrectly adjusted. Zorskie felt the whole thing was scientifically embarrassing and realized how uncritical Bohm could be with new people and new ideas.

Holding strong reservations about the alleged phenomenon, Hiley pointed out to Bohm the danger, for their other work, of appearing to endorse Geller. Finally Bohm agreed to back away. In many ways Bohm's naïveté was endearing for, if time permitted, he was always willing to talk to people and was open to ideas that others would readily dismiss. But the result was, as Hiley put it, that Bohm often had to be saved from idiots![12]

Through his books and essays, Bohm's work was attracting considerable attention in the United States. Here was a thinker of distinction who could provide a scientific and philosophical basis for new ideas that were emerging in education, religion, and psychology. Soon Bohm was a regular traveler, visiting the United States twice a year to give talks and spend time with different organizations.

Nineteen seventy-nine was the year of his first visit to the U.S. since his father's illness. His destination was Ojai, a small town an hour's drive north of Los Angeles, that had been chosen several decades earlier by Krishnamurti for its aesthetic and spiritual qualities. From Ojai he could visit Clairmont College and talk with the philosophers John Cobb and David Griffin, who were studying Whitehead's process philosophy. Rumor had it that Bohm would join the college, but he decided that with so much emphasis on Whitehead, he would not be entirely free to pursue his own ideas.

California gave him the chance to renew contact with Richard Feynman. On one occasion Bohm arranged a lunch with Feynman and Krishnamurti, at which Feynman talked about his childhood and his relationship to his father, then went on to say that physics was stuck because of a lack of imagination. When he mentioned his own lack of interest in the philosophical issues of science, one of the Ojai group, David Moody, joked, "Dave knows a little bit about both." Feynman became angry, saying, "I can tell you one thing. David Bohm knows a lot more than just a little about physics."[13] Booth Harris, a teacher at the Ojai school, remembered Feynman saying, "You probably don't know how great he is," and noticed the considerable respect Feynman showed toward Bohm.[14]

Ojai was the center of another of Krishnamurti's schools, and soon Bohm was playing a significant role in the school's life, as he had in Brockwood Park. Booth Harris had been drawn to Ojai not so much by Krishnamurti as by the ideas of David Bohm, who was a great influence on him. Harris connected with Bohm in a human way, feeling him to be a person who had experienced great suffering. He believed that Bohm's concern with wholeness arose from his deep need for connection to the world. Above all, it was when Bohm was with Krishnamurti that he felt a sense of immediate connection.[15] Lee Nichol, another teacher at Ojai, found Bohm was "saturated in the spiritual," which left him particularly vulnerable to Krishnamurti's influence.[16]

Nichol soon drew close to Bohm. A child of the 1960s, Nichol had an attitude toward authority and discipline that left him confused in his role as a teacher. One day Bohm suggested, "You must tell children how to behave, but you mustn't tell them how to be." That remark transformed Nichol's approach and, at the same time, illustrated Bohm's idea of the difference between a paradox and a contradiction: A contradiction involves two things that cannot fit together, while a paradox, which appears at first sight to be a contradiction, on closer examination has a resolution.

Nichol wondered if Bohm was capable of entering the same states of consciousness as Krishnamurti. He once asked the physicist if in his discussions about Krishnamurti he was merely talking about a particular transformed state of consciousness, or if he had actually experienced it. "It comes very often," Bohm replied, "but it's fragile. It's not solid or consistent in its feeling." When pressed, Bohm said that being the object of other people's attention was the strongest factor in bringing him out of that particular state of consciousness—a state which presumably was close to that transformation referred to by Krishnamurti.[17]

A few years earlier Bohm had told me something along the same lines. In 1974, when he visited me in Ottawa, we spent many days talking together about Krishnamurti's teaching. At the time I challenged him, in the amiable spirit of our discussion, that unless he too had experienced the state Krishnamurti described, then his discussion of transformation was hypothetical and empty. To this Bohm replied, "Well, let's say I have seen some of the things Krishnamurti talks about. I have looked at reality and seen that it is an illusion."

Many have wondered about the extent of Bohm's involvement with the Indian teacher. Was his understanding of Krishnamurti purely "intellectual," or had he too entered into a state of transformation? But the statements Bohm made to Nichol and myself (and there may have been others) are unequivocal. Perhaps of more significance than the possible objectivity of these accounts is Bohm's own belief on those two occasions that he had undergone some form of inner transformation.

In the following year, 1980, Bohm returned to Ojai and took the opportunity to drop in on his old friend Mort Weiss, who lived in Laguna Beach, south of Los Angeles. The two men reminisced about the last time they had been together in the early 1950s, when Weiss had said good-bye to his friend at Pennsylvania Station. Recalling Bohm's passion for walking, Weiss took him to the top of the hill behind his house, where there was a view of the adjoining valley.[18]

As they walked, Bohm experienced a severe pain in his chest. Weiss, who had heart problems himself, offered Bohm one of the nitroglycerin tablets he carried. The pill worked to the extent that Bohm was able to make his way back to Weiss's home, but there he collapsed into a chair, breathing heavily and barely able to speak.

Weiss urged him to go to the hospital, but Bohm had no proper medical insurance and could not afford the fees. In the end Weiss drove him to his own heart specialist, who advised immediate admission into the hospital. Bohm became so worried about the costs that the specialist prescribed medication that would enable him to get back to England. Weiss then drove Saral and David to Los Angeles airport, where they caught a flight to Heathrow.

Back in London, medical tests at Edgware General Hospital indicated that the whole incident had been nothing more than a digestive problem. Only when Bohm again became ill, this time at Birkbeck College, was he sent to the University College Hospital. At first sight his EKG scans again looked normal, but abnormalities were discovered upon closer examination, and Bohm was scheduled for an angiogram at Bromptom Hospital. The test showed that two coronary arteries were totally blocked and a third almost blocked. Almost a year of misdiagnoses had passed since his first chest pains in Los Angeles.[19]

Bohm's condition was serious. Without a triple bypass, he was unlikely to live for another year. There was, however, a complication: The delicate nature of Bohm's arteries would make the graft difficult. The consultant gave the operation only a fifty percent chance of success.

Hearing of Bohm's emergency condition, Krishnamurti wanted to attempt a healing before he went into hospital. He suggested they try it over lunch at Fortnum and Mason. The location proved too bizarre for Bohm, and they compromised on his office at Birkbeck. Saral and Mary Zimbalist waited outside. Bohm did not speak of it later, but Saral deduced that it involved some form of laying on of hands.

The bypass was scheduled for June 25, 1981, and Bohm entered the Brompton Heart and Lung Hospital as a National Health patient. His cardiologist was Dr. Honey and the surgeon in charge of the procedure Mr. Paneth. Given a choice between a single room and one with four beds, Bohm chose to stay with others. Saral noticed that he seemed at home with these nonacademic men, who took him under their wing.*

In order to be closer to the hospital, Saral stayed with her cousin,

* Britain operates a two-tier medical system. A person may be treated under the government-administered system, as a National Health patient, or privately, generally under the auspices of private health insurance. While there are privately operated clinics, private patients can also obtain treatment in National Health hospitals, availing themselves of private nursing and shorter waiting periods for treatment.

on the other side of Hyde Park, the night before the operation. Bohm's operation was scheduled for eight in the morning and was expected to last four and a half hours. At noon Saral telephoned the hospital and was told to call back in half an hour.

As she and her cousin prepared lunch, the telephone rang. The hospital asked Saral to come immediately. She jumped into a cab, only to find traffic blocked at Hyde Park corner. When she explained the emergency, a policemen switched on his siren and escorted them through stoplights and traffic.

Upon arriving at the hospital, Saral learned that Bohm was back in the operating theater. Following the bypass, his blood pressure had collapsed, and they had had difficulty maintaining his vital signs. His heart had been damaged. He was rushed to a clean operating room and reconnected to a heart-lung machine while the doctors worked for another four hours to revive him. Bohm remained in a coma for two and a half days. During that time the staff suggested that, even though he was unconscious, Saral should sit and talk to him.

Krishnamurti, also concerned about Bohm's prognosis, had asked Saral to inform him of the exact time of the operation and to tell Bohm that "I love him."

When Bohm recovered consciousness, Saral asked him if he had had any sensations. He said that he had had a sense that he could easily have let go and died. The prospect had not seemed too bad, but he then had realized that he still had work to do and made himself come back.

As the days went on, Bohm improved and Saral asked the surgeon whether he was cured and could now have a normal lifespan. To her surprise he told her that the problem would quite likely recur within five or six years. As it turned out, Bohm was lucky; he had another twelve creative years before his heart finally caught up with him.

Although his heart was damaged, he was soon able to resume his long daily walks. Still a degree of fear remained, and he was constantly monitoring his pulse and acting like someone who is never too sure that his very old car is going to make it around the next corner. His obsession with his pulse became something of a joke among his friends. Hiley would ask, a serious expression on his face, "Is it still beating, Dave?"[20]

The following year he was well enough to travel again, and he began the series of trips that were to bring his ideas to an even wider audience. In the United States the Kettering Foundation organized a conference around Bohm and Owen Barfield. Barfield, a poet and solicitor, had been a member of the distinguished Oxford group the Inklings, whose members included J. R. R. Tolkien and C. S. Lewis. In addition to his short but illuminating books such as *Poetic Diction* and *Speaker's Meaning*,[21] Barfield had produced a major study of Coleridge. Prior to the conference Bohm had had a number of meetings with Barfield at the prestigious Athenaeum Club in London. He found the contact particularly stimulating and liked to discuss Coleridge's distinction between "fancy" and "imagination." He also found congenial the poet's dialectical oppositions, such as that between reason and understanding, and his stress on "organic unity."

In the fall of 1982 Bohm was at Syracuse University, in New York State, as the Jeannette K. Watson Distinguished Visiting Professor—an award that in other years had been given to Saul Bellow and Noam Chomsky and that later went to Ernst Gombrich. The amount of work Bohm agreed to do would have been considered excessive even for a younger man. From September 13 until October 1 he gave twelve lectures in the Watson series, each one carefully thought out and written down. In addition he gave public lectures on Monday evenings, Tuesday afternoon seminars, and a variety of additional seminars and discussions on physics, language, and art in various departments of the university.

In Syracuse Bohm and I took walks together and planned a series of essays on art and science. The project eventually was transformed into the book *Science, Order and Creativity*.[22] Instead of accompanying us, Saral remained in the apartment, as she was troubled with pain in her hip following a fall earlier in the year. Her spirits were becoming even lower. A year ago her husband had nearly died, and now he was facing retirement. Both were concerned about how they would make ends meet. Since Bohm had been middle-aged when he had started teaching in England, he would not receive a full retirement pension.

Saral's solution was that they should move to Israel, where she had friends and relatives—a prospect far from welcome for Bohm.

Syracuse gave Bohm the chance to see Joe Weinberg, the friend from Oppenheimer days who was now head of the physics department. It turned out that Bohm's seminar to the physics department did not meet with Weinberg's approval. It would cause students to question the authority of their professors, Weinberg told him—a remark Bohm found difficult to square with the atmosphere that had once prevailed at Berkeley.[23] At the dinner afterward Saral watched her husband laughing and talking at the other end of the table. Weinberg said to her, "It's terrible what's happening with Dave and Krishnamurti." Saral asked him what he meant. "He's deserted physics," Weinberg replied. Saral said that he hadn't really left physics, and anyway he was happy. Weinberg repeated the word, "Happy?" as if he felt happiness was not particularly important.[24]

The following year Bohm attended a conference at Carleton University, in Ottawa, that had been organized in his honor. There he talked and responded to panels of artists, musicians, poets, physicists, and philosophers. At the Science Council of Canada, a policy advisory group to the Canadian government, he discussed the complex chemical ecology of the brain. Subtle chemical changes color our subjective thinking process, he said, and in turn our thinking induces objective changes in this chemistry. A man walking at night sees a potential assailant. The sight produces perturbations in his brain's electrochemical activity, which in turn profoundly influence his emotions and disposition for action. Yet as soon as he realizes that the assailant is nothing more than the shadow of a tree, his entire brain chemistry changes. As Bohm later came to put it on many occasions, "A change of meaning is a change of being."

These ideas were worked out more fully in a series of drafts Bohm circulated over the next years: "Soma-significance and Signasomatics" and "A New Relationship between Mind and Matter."[25] Words, he said, unfold in the brain, producing changes in its chemistry that permeate the whole body. As a result of our individual and social conditioning, particular words evoke strong somatic reactions that in turn modify our thinking. Words, thoughts, feelings, and intentions have their objective correlates as chemical processes within the brain;

likewise, objective chemical processes have their subjective correlates in movements of thought. Thought is neither exclusively subjective nor exclusively objective. Like the observer and the observed in quantum theory, or the poles of a magnet, the two are inseparable. "Significance" always has a somatic component, and somatic changes are always accompanied by a change in mental significance.

In this period of his life, despite his recent life-threatening condition, Bohm's ideas fed each other—mind and matter, implicate order, causal interpretation, and prespace algebras. He was now talking about a new superimplicate order. In *Wholeness and the Implicate Order* he had proposed that our everyday world unfolds out of the underlying Implicate Order. The depths of the implicate can never be totally unfolded; it is the source of the explicate order yet remains unconditioned by it.

In his discussions with Bohm, the biologist Rupert Sheldrake objected that this duality sounded a little like a reformulation of the Platonic ideals.[26] A particular chair, the explicate, is the manifestation of an ideal chair (the implicate) that has no material existence. Others found the relationship between implicate and explicate order to be curiously one-sided. The implicate chooses to manifest itself in the theater of the explicate, yet it does not witness its own display. An explicate drama is played out in the world of space and time, yet it is never carried back to inform the implicate.

Bohm realized that a further level was required. The superimplicate order, as he called it, lies below the implicate. In keeping with his earlier vision of a universe made up of an infinity of levels, he left open the question of additional levels of enfoldment lying beneath the superimplicate. When the explicate unfolds out of the implicate, its temporal and spatial manifestation is witnessed by the superimplicate. In turn this superimplicate acts on the implicate order. The result is feedback loops similar to those found in a video game. (The image of the video game was suggested to Bohm by Alex Comfort, a gerontologist with a serious interest in the mathematics of prespace but better known to the general public as author of *The Joy of Sex*. He offered an analysis of Bohm and Jung in *Reality and Empathy*.[27])

Suppose you shoot down spaceships on a video game. These spaceships have no material existence; they are only moving patterns of light on a television screen. The display itself is a manifestation of the underlying program within the computer. The ships are analogous to the explicate order. They are a surface appearance that subsists on the underlying program, analogous to the implicate order.

And what of the game player? He or she responds not to the underlying program but to the appearance of the ships on the screen. The operator takes actions based on a perception of the explicate order and, as buttons are pressed, they affect the underlying program, the implicate order. Clearly the human operator plays the role of the superimplicate order.

The new idea was inextricably linked to his revived interest in the causal interpretation. In the original version of this theory, the electron is a real particle moving along a definite path determined by the quantum potential. Back in his early days at Berkeley, after working on elementary particle-scattering problems in which the scattered particle spreads out like a wave, Bohm had conceived of particles in terms of processes. A particle is a wave that alternately collapses inward from the whole universe and then expands outward. Yet his hidden variable or causal interpretation pictured the electron as an actual particle that is guided by the quantum potential. But what about the electromagnetic field that conventional quantum theory pictures as composed of localized photons? Bohm and Hiley realized that the light (and other fields) must be treated as pure fields so that the photon becomes a global or nonlocal object. (For the technically minded, the photon is a nonlinear effect of the entire field.)

In conventional quantum theory it is impossible to give an underlying explanation of how an excited atom emits a photon of light. Bohm and Hiley now pictured the electromagnetic field as sweeping inward toward the excited atom, then expanding outward again, taking with it the atom's excess energy. In what is known technically as the second quantized version of this theory, Bohm showed that these field processes are guided by a new super quantum potential. (Of course, for Bohm, this idea was simply another level of explanation. He believed that there would be other, qualitatively different levels, mutually supporting each other.) Clearly there are parallels between the superimplicate order and the super quantum potential, which Bohm pictured using the image of a video game.

The metaphor of the video game was also applied to Bohm's ideas of soma-significance. As we play the game, those spaceships on the screen seem so real to us that we forget that they are merely the manifestation of an underlying program. Similarly, the mind offers a "display" of its own—anger, fear, desire—mental images that appear to be so autonomous and independent of us that they become other, something that the "I" witnesses and attempts to control. Yet there really does not exist an "I" that is absolutely separate from these images. As Krishnamurti had said before, the thinker *is* the thought.

The whole mental "display," Bohm argued, is a subtle trap. What we are experiencing is merely the "weather" of the brain. Just as we watch the wind and rain around us without feeling responsible for it or trying to change it, so too we should observe the brain's "weather" and realize that it is an electrochemical process generated by memory and previous conditioning. But this separation is enormously difficult to see. The reason we mistake mental processes for part of ourselves is that our brain has no nerves that register its own thinking. We have no way of perceiving how the display operates, no senses that convey the origin and dynamics of thought.

Here Bohm drew upon the notion of proprioception—our ability to know exactly where our arms and legs are situated in space, even with our eyes shut. Subtle feedback signals from our muscles constantly inform the brain of the body's disposition so that, for example, we always know what our hands are doing. But we have nothing that tells us "how" we are thinking. All we have is the product—individual thoughts, feelings, and intentions. This lack of proprioception for thought is the basic flaw in the human animal, the origin of the mental trap.

How can consciousness display itself? Bohm now asked himself. How can we get out of the trap? How can the mind develop its own proprioception?

CHAPTER 15

Dialogue and Disorder

BOHM WAS NOW a regular traveler to the United States, but since he entered on a visitor's visa, he could not obtain honoraria for his lectures. He decided to look more seriously into obtaining a visa that would permit him to accept fees. He visited Edward S. Gudeon, an American immigration and citizenship lawyer working in London, who took on the case *pro bono publico*—that is, without a fee.[1]

The original confiscation of his passport in Brazil had not amounted to a loss of American citizenship. This had occurred later, under amendments to the 1952 Immigration Nationality Act, when Bohm became a naturalized Brazilian and took an oath of allegiance to that state. Gudeon advised Bohm that subsequent Supreme Court decisions had opened the possibility of attacking this loss of citizenship and reinstating Bohm.

Bohm's Brazilian naturalization, Gudeon argued in a brief to the U.S. State Department, was an involuntary act done under extreme duress. At that time, he said, the physicist had not intended to relinquish his U.S. citizenship. Unable to travel abroad, Bohm had been prevented from furthering his scientific career by attending conferences, giving talks, and meeting other scientists. "Physics has occupied virtually all of Dr. Bohm's working life," he argued. It was "the principal source of meaning to his life and the sole driving force which has motivated and continues to motivate his daily life and actions."

Bohm added a statement describing his naturalization: "At the time of my naturalization, I was one of a group of people to whom an oath of allegiance to Brazil was administered collectively. We all indicated assent either verbally or else it was implied that if we said nothing, this signified assent. The oath contained a renunciation of all previous nationality. I recall that my feelings at the time were of surprise, but never disloyalty. I did not wish to give up my US citizenship and I certainly did not feel disloyal in any way." The file also included correspondence with Einstein that testified to Bohm's state of mind at the time.[2]

The State Department accepted the lawyer's argument, and Bohm's citizenship was restored *nunc pro tunc*—back to the beginning. Legally speaking, Bohm had always been a U.S. citizen. Learning this was particularly important to him, as he told David Shainberg, since he had felt "thrown out" of his country and rejected.[3]

With his citizenship restored, Bohm's friends and colleagues explored possible positions for him in the United States. Yet despite his constant worry about money, Bohm never seriously considered permanently returning to his native country. He was, however, amused at one proposal—that he should become a professor of creativity!

In Oxford the philosopher of science Harvey Brown suggested a fellowship at Wolfson College. The Bohms drove down and looked at possible accommodations, but David did not want to be so far away from Basil Hiley.[4] While he was walking around one of the older Oxford colleges and looking at the paintings in the dining hall, he remarked, "Too much tradition."

Retirement did not diminish the intensity of Bohm's work, but it did mean less money. Another nagging concern was Saral. Her hip problem had cleared up, but she was still in low spirits and dissatisfied with life in Edgware. A nearby friend was moving away, which would leave her even more isolated. She often thought of Israel, and now that her husband had retired, she felt they had no need to stay on in London. All these feelings led to endless discussions.

In November 1983 I visited Bohm in London to work on our book. Collaboration was difficult since Bohm was quite distressed. During our walks he could not concentrate on physics but went over and over the question of whether they should move. The prospect produced great tumult in his mind, yet he acknowledged that Saral was fed up with Edgware. The tension at home had reached the point where it was difficult for him to work. What would happen if Saral's depression continued? He couldn't stand it. What if it got so bad that he could no longer think? Should he leave her? But how could he live on his own? His agitation was extreme.

He had other worries, too, that I did not know about at that time. While he remained convinced of the truth of Krishnamurti's teachings, he was increasingly troubled by developments at the Ojai and Brockwood Park schools. Due to tensions and resignations, both seemed to be falling apart.[5] Krishnamurti, he felt, did not give sufficient attention to the social dimension in his teachings. Those who surrounded Krishnamurti were like spokes on a wheel, related to the center, which was Krishnamurti, but not to each other.

Also disturbing was the way Krishnamurti's image was being inflated by those around him. The Indian teacher seemed to do nothing to prevent it. As far as Bohm was concerned, the very idea went against everything Krishnamurti had taught. Was it possible that even he had been conditioned by his curious upbringing? Yet how could he be? Surely Krishnamurti was beyond conditioning. It was with a troubled mind that, in the spring of 1984, Bohm flew to Ojai to meet with Krishnamurti once more.

Another context surrounded their Ojai meeting, although the evidence for it is largely based upon gossip and therefore difficult to verify. Bohm's closeness to the Indian teacher, together with his influence on the manner of Krishnamurti's presentation, had always caused concern among some of those around the teacher. Bohm's increasing importance as an international speaker was also feared, for should the Indian teacher die, people would assume that the way to Krishnamurti's teachings lay though David Bohm. The physicist might even be taken as Krishnamurti's successor.[6]

Perhaps the concerns about Bohm's role influenced Krishnamurti himself. In retrospect it is hard to determine, but it is certainly true that while some welcomed the closeness between Bohm and Krishnamurti, others would have been happy to see a wedge driven between them. Their first meeting at Ojai appeared to go well at first, but then Krishnamurti began to push Bohm to consider the nature of his ego. Increasingly Bohm felt himself forced into a position of nothingness.[7]

The underlying theme of Krishnamurti's questions was Bohm's dependence—not only on Saral but on the teacher himself. How was it, Krishnamurti wondered, that he and Bohm had talked together for so many years, and in such an intense way, yet nothing in Bohm's nature had fundamentally changed? Had Bohm responded only at an intellectual, superficial level? If Bohm, with all his energy and clarity of mind, was not truly touched by the teachings, was there hope that anyone else could be?

As Krishnamurti confronted Bohm in a way that others later described as "brutal," the physicist was thrown into despair. Unable to sleep, obsessed with thoughts, he constantly paced the room to the point where he thought of suicide. At one point he believed that he could feel the neurotransmitters firing in his brain—a condition he related to his insights on the brain's chemical ecology. His despair soon reached the point where he was placed on antidepressants.[8]

As Bohm's condition worsened, Krishnamurti distanced himself from the physicist. Bohm's breakdown confirmed his suspicion that his teachings had not permeated the physicist's life in any fundamental way. He no longer felt able to explore ideas with Bohm as they had done in the past, and while he remained cordial, the intensity between them vanished.*

Krishnamurti's reaction raises a troubling question. If he believed that Bohm had understood his teachings in only a superficial fashion, then what did he think had been taking place between them for all those years? Their discussions had been exceptionally intense. Krishnamurti had spoken of something remarkable in their presence together, of the arising and dying away of thought, and of immense clarity. At such moments Krishnamurti had genuinely believed that Bohm was participating in the same process as he. How was it possible for a mere "intellectual" relationship to have created such a deep sense of understanding between the two men?

Back in London Bohm began regular treatment with Patrick de Mare, a Freudian analyst.[9] He now had another source of distress—

* Scott Forbes, who had been a teacher at Brockwood Park, spoke to Krishnamurti about what had happened. He was told that Krishnamurti no longer felt he could explore with Bohm to the same extent as before. This explanation was confirmed by Booth Harris, Mark Lee, and Lee Nichol at Ojai.

his teeth, requiring dental work that proved very expensive. Bohm connected his defective teeth to early childhood traumas. Early in his marriage he had sometimes awakened in the night in shock and distress, crying out, "My teeth, teeth, teeth!" Teeth he associated with moral and spiritual strength. The man who had once been called "sugar freak" now had no teeth in the front of his mouth and wore a bridge that he removed at night. The remainder of his teeth were in poor condition.

Yet out of the depths of his distress, Bohm found a new strength. In addition to conventional Freudian analysis, de Mare worked with groups—not the "T" or encounter groups popular in the late 1960s but groups as a form of social therapy. De Mare believed that in the hunter-gatherer stages of human social development, when people had lived and traveled in groups of thirty to forty, social and psychological tensions had been dealt with as they arose, through a process of dialogue. It was after the growth in size and complexity of human societies, following the transition from farming to city building and finally to industrialization, that the power of the group disappeared. But human beings are not psychologically well adapted for life in complex societies, de Mare believed, and require continuous, active social therapy.

De Mare's ideas appealed to Bohm, who believed that each human being contains three dimensions: the individual, the social, and the cosmic or religious. Krishnamurti addressed two of them but ignored the social; the social dimension had always been important to Bohm. Once he had believed that social transformation would come about through Marxism. While giving his seminar at Berkeley, he had experienced an almost mystical sense of group consciousness. In his physics he dreamed of doing research that was noncompetitive and truly collaborative, that would bring scientists together through bonds of impersonal friendship.

Not only did he attend de Mare's group for his patients, but with his friend Maurice Wilkins, he joined a "theoretical group" run for therapists. In these groups Bohm found a new way of synthesizing his ideas on consciousness. To take a hypothetical example: Suppose you are talking to someone who makes a remark that you find idiotic, prejudiced, or just plain wrong. You try to set them right, only to get an antagonistic response. Although you try to remain calm and rational, pretty soon you have been dragged into a heated argument. This process is even more disastrous when it occurs between nations instead of between individuals. The source of these tensions, Bohm had earlier suggested, lies in a basic flaw in human nature—the absence of a display for the action of thinking, and the very speed with which thought reacts to itself.

Dialogue groups, Bohm realized, were a way of slowing down the thinking process and displaying it in a public arena. They held the solution he had been seeking for years: a way of bringing about a radical transformation of human consciousness. When two people argue, it is often because the words used by one trigger off a series of complicated internal reactions in the other—causing a change in the brain's "weather," as he put it. This complex movement of electrochemicals conditions further thoughts, which in turn produce more "weather." Even when we believe we are behaving reasonably, our thinking is still trapped in its own chemistry. The essential problem is that the process itself happens so very quickly that we do not notice the game between impulse and response.

An analogy may help. Uranium disintegrates very slowly. An individual uranium nucleus can be around for thousands of years before it splits. When it does so, it releases neutrons. Should one of these neutrons happen to hit another uranium nucleus, it will cause it to disintegrate immediately. Atomic reactions require the controlled disintegration of uranium, but these neutrons travel so fast that they escape from the atomic pile before they have a chance to initiate further disintegrations. To overcome this problem, scientists use what are called "moderators"—graphite or heavy water—that act to slow down the neutrons. Slow neutrons spend much more time in the atomic pile and will eventually collide with a uranium nucleus, causing it to disintegrate and emit more neutrons. In this way a sustained chain reaction is possible.

Words of totality, such as *never, always,* and *totally,* may be mental "neutrons." They reflect people's nonnegotiable positions and their political and religious conditioning. (These words, it will be recalled, were frequently used by Krishnamurti.) The dialogue group, Bohm believed, slows the processes of thought down to such an extent that the effects of "neutron" words can be observed by others. When two people in the group argue, it becomes apparent to the others that both are operating from fixed conditioning deep within their natures. In a group of thirty to forty people, there will always be some who do not hold either extreme position and can help to defuse an argument. They do so not by striking a compromise but by discovering new ground where the two opponents may hold their positions in tension until a possible resolution is reached. Moreover, by displaying anger, aggression, confusion, fear, pain, and so on within a group situation, people may experience these emotions in a spirit of dispassionate friendship. If the group meets sufficiently often, Bohm believed, and if its members are truly serious, then something akin to a group mind will develop.

Bohm had long wrestled with the issue of the individual and collective. His first significant research had been on the plasma, in which, as a reflection of Marxist philosophy, he had seen the freedom of the individual as arising out of the collective, and the collective as enfolded within the individual. His reformulation of the EPR paradox, which led to Bell's theorem, demonstrated that even widely separated objects can be intimately correlated—connected not through any field or interaction but through an indissoluble, acausal link. Finally, his causal interpretation had pictured individual electrons as responding to a collective pool of information.

Could not a similar process operate within a social group, he wondered, one in which each member generates intensity sufficient to transcend the normal limitations of individual consciousness? If so, it might be possible for consciousness to function in a truly collective way, with each speaker responding to a common pool of information. Some years earlier Bohm had speculated that simply being in the presence of Krishnamurti and engaging in an active inquiry was sufficient to induce a transformation of consciousness. Even if only a handful of persons could participate in this transformation, he had thought, they would still be sufficient to induce a radical change in general human consciousness. Now he believed that such change could come about through the operation of group dialogue. Not only would the group respond to a collective pool of information, but its effect would extend beyond the boundaries of the group itself, inducing subtle but significant effects within human society.

In dialogue people were made aware of the difference between thought and thinking. Thinking, for Bohm, is an active movement, while thoughts are fixed forms based on responses to memory and the past. Analogously, Bohm suggested, there was the difference between feeling and "felts." "Felts," like thoughts, are fixed forms based on reactions to past feelings. When people say that they "trust their feelings," Bohm was quick to point out that they were generally operating not from feeling itself but from "felts."

Dialogue works at several levels. At the deepest level it is about the development and transformative power of the collective mind. At another it provides a "display" of thought, slowing down its movement and allowing its observation. It allows the expression of many alternative views on a particular topic, some of which are presented in nonnegotiable ways. Thanks to the group process these differences do not lead to direct confrontation but are held together in a creative tension. Rather than trying to resolve opposing positions through compromise, it is possible to move to an "order between and beyond."

In science an evolutionary model applies to the way rival theories compete until one replaces the other. Bohm believed another movement was possible, one in which different theories and approaches coexist and illuminate an area of knowledge in a more creative way. Schrödinger's wave mechanics, he pointed out, was discovered a year after Heisenberg's quantum mechanics.* Initially Schrödinger believed that he had created a radically different viewpoint on the quantum world, but Bohr was quick to convince him that his own approach was no more than a mathematical transformation of Heisenberg's. Bohm believed that the two versions are indeed subtly different and that they should have been held together, their similarities and differences emphasized, for much longer so that something new and qualitatively different could emerge from their tension. Likewise, the initial confusion about the meaning of quantum theory

* Schrödinger later demonstrated the formal equivalence of his own and Heisenberg's approaches.

was resolved at Copenhagen far too rapidly for Bohm's liking. It did not allow for a buildup of creative energy that could have produced something new. His own ontological interpretation should coexist in tension with conventional quantum theory, so that an "order beyond" both theories could be discovered.

Such plurality is anathema to most scientists. Yet it is becoming the norm in such social areas as conflict resolution and negotiation. In *Conversations Before the End of Time,*[10] Suzi Gablik explores questions of art and the environment through a series of dialogues, specifically acknowledging David Bohm, Arnold Mindel, James Hillman, and Michael Ventura for illuminating the process. Just as artists once discovered collage, she argues, so dialogue juxtaposes different, and strongly held, views on a subject.* Rupert Sheldrake attempted to do something analogous. By taping and then transcribing discussions, he hoped that something would emerge that was different from the more conventional, monolithic viewpoint of a single author.

The ability to support unresolved paradoxes and to allow many different styles and interior dialogues to flourish is the mark of a truly creative scientist, artist, writer, or musician. While he may have been politically conservative, Shakespeare presented in his plays an entire universe of widely differing personae, each with his or her unique voice. For Bohm the individual is enfolded within the social and the social within the individual. People with sufficient creative energy can, by working on their own, dissolve fixed thought and provide the fertile ground to sustain a multiplicity of voices. Yet most of us normally use our energy to sustain a false sense of ourselves, which means we tend to operate from fixed and nonnegotiable but unexamined positions. Here lies the power of dialogue: to make manifest such assumptions and positions, bringing them out into the open.

Bohm was eager to try out these ideas and create a dialogue group of his own. It was to be a microcosm of society at large. By holding the tension between different positions for much longer than is normally possible, the group would generate the very "display" that consciousness normally lacks. A person's urgent need to say some-

* Gablik's point is that our society must move from considering the artist to be a unique hero into realizing that each of us is a special artist who can touch the sacred.

thing might, for example, be suspended, so that he or she could observe the way intention operates within the mind and body. Another, who normally remains silent, might be moved to speak, making it possible to observe the processes of thought within the body.[11]

Dialogue groups now occupied an increasingly important place in Bohm's life. Here was an approach that would genuinely help transform society, he believed. A welcome opportunity to extend this process occurred during a visit to Mickleton.

Bohm had earlier been approached by two young men, Don Factor and Peter Garret, in connection with a conference they were organizing.[12] The result of their discussions with Bohm was to suggest an informal weekend for a group of friends and other interested people in the village of Mickleton in the south of England. At first Bohm was very nervous and read from a prepared talk, but soon he found a spirit of friendship developing in the group, and rather than continuing to read from his prepared talk he decided to move into an active dialogue that would involve all the participants.

At Mickleton he explained that a change of meaning is a change of being. Meaning is usually taken to be subjective, Bohm explained, but thoughts and words play an objective role in modifying brain chemistry. When a change of meaning is profound, it brings about ontological transformations in the individual, specifically a subtle but permanent restructuring of the brain. Such transformations can actually take place as a result of group dialogue, he believed.*

In the next years Bohm organized dialogue groups in a variety of locations—Israel, Geneva, Sweden, Denmark, and the United States. A young Canadian, Mario Cayer, visited the London dialogues, then set up his own group in Quebec. Bohm's friend David Shainberg hosted a regular group in New York City.

Most of Bohm's dialogue groups had met on weekends, but now he was ready for a more sustained experiment. Roger and Joan Evans, of the psychosynthesis movement, were representatives of a foundation that had given an award to Bohm. Over dinner, they asked him if

* One of these weekend groups was transcribed into a book, *Unfolding Meaning*. It appeared to irritate the physicist John Taylor, who wrote a review in the *New Scientist,* accusing Bohm of becoming a guru of science.

they could help him with this work in any other way. His answer was "Dialogue." So it was that a group began to meet every two weeks in the Evans's home at Mill Hill, in north London.[13]

As a psychotherapist, Joan Evans had had experience in group work and knew that human groups pass through various stages of anxiety, conflict, search for leadership, and cooperative integration. As Bohm explained that he wanted to take group dialogue into a new dimension, she was interested to see for herself how he would put this idea into practice. During the first meetings of the group, Bohm described his approach, giving a strong sense of vision and of the goal he was reaching for. Yet in Evans's opinion he did not seem to know concretely how to make this happen.

Earlier, Peter Garret had also found Bohm to be particularly articulate when he introduced participants to his dialogue process. Yet in the group his own charismatic presence had an overawing effect, so that most of the participants deferred to him. The very speed of his thinking, as well as its subtlety, made them feel bulldozed.

Another difficulty had been noticed independently by Garret, Evans, and Shainberg: the degree of Bohm's control over the group.[14] Dialogue, in Bohm's terms, should be truly leaderless, even if a facilitator is needed in the early stages. Yet in Evans's experience any group goes through an initial period of looking for a leader. Bohm himself appeared highly ambivalent toward his own authority: On the one hand he would not take responsibility for his own role as originator and help the group through this stage; yet on the other he appeared, unconsciously perhaps, unable to give up control. All three noticed the subtle body gestures, intakes of breath, and hand movements whereby Bohm expressed encouragement or disapproval. At times he became conscious of this effect and attempted to distance himself by sitting in the outer circle and asking someone else to act as facilitator.

It was for these reasons that Evans, at least, felt that the whole enterprise was a failure. Admittedly the fault also lay with the participants, who were attending mainly because they wanted to be with Bohm, never entering into the spirit of a genuine quest. But Garret was not deterred and continued working with dialogue groups even after Bohm's death. Today many dialogue groups, looking toward Bohm's initial impetus, are operating all over the world. Another group, convened by Don Factor, has been meeting regularly in London for many years.

To criticize the application of Bohm's method is not to deny its validity. Dialogue is fundamentally an experiment, and like all experiments, it can have positive and negative results and different experimental arrangements. As to what the groups achieved, opinion is strongly divided. Some participants feel that dialogue groups should continue even in Bohm's absence; they are an extremely important nucleus for bringing about social transformation. Others concluded that the process would never work as Bohm hoped, or that the experiment should move in a variety of different ways. Still others argue that group dialogue should be adapted to immediate practical ends—used as a management technique, for example, or to facilitate the investigation of practical problems. Needless to say, this use was anathema to Bohm. His dialogue groups were without any goal, for as soon as a goal is established, he argued, then freedom to explore is limited.

Science can be thought of as a dialogue with the natural world. In Bohm's case, science-as-dialogue was a matter of remaining open not only to external nature but to one's inner psychophysical promptings. While he did much of his physics in solitary contemplation, he also enjoyed working with other people and, above all, talking with people while he walked. Science, he believed, should have a social dimension. In the mid-1970s, at a conference at the University of Western Ontario, he had suggested that the participants abandon the traditional structure of podium and rows of chairs and sit in a circle. While this practice is commonplace today, it was certainly unnerving to some of the physicists and philosophers present.[15]

Bohm had also organized a series of circles in which scientists could meet with Krishnamurti. While their formal interactions were not that successful, Bohm discovered that, during lunch breaks, the participants talked more openly about their inner motivations.[16] At a conference on theoretical biology, organized by C. H. Waddington in Bellagio, Italy, Bohm suggested that instead of presenting and defending particular theories about aspects of biology, it would be better if the participants spoke about the metaphysics and philosophical assumptions that underlay their respective approaches. Bohm's proposal was an eye-opener, but his suggestion was taken up. For many it was the first time they had spoken openly about metaphysics.[17] His dream was to organize a true scientists' dialogue in which a larger group would come together to talk not about their theories but about the nature of science as such and about their belief-systems. From time to time he would make plans for such a gathering, but it never materialized.

In a sense dialogue was the method whereby Bohm and I worked together on *Science, Order and Creativity*. The idea of the book had emerged out of our many discussions. When we finally met, once or twice a year, to write the book, our method was always the same. Over morning tea at the Bohms' Edgware home, we would read over the material that we had written the day before, then take a walk to talk over new ideas or search for an alternative approach. This discussion would continue over lunch until it was time for Bohm's afternoon rest.

For the remainder of the afternoon, we would set down on paper our conclusions for the day. When Bohm was fully in the swing of talking, it was never a good idea to interrupt him, even when he seemed to be going in the wrong direction. In the grip of a global idea, he could make sweeping statements that on closer examination appeared nonsensical or contradictory. He saw things in a gestalt and needed to unfold his ideas as fast as possible before that vision was lost. Interruption irritated him because it arrested the momentum of his thinking. It did not much matter whether a particular idea made sense, since he was sketching out an overall pattern of ideas. As a result he was quite capable of shifting and even reversing his position, or making contradictory statements.

Roger Penrose once remarked to me that Bohm was like the quantum mechanical wave function itself. As he spoke, he would spread out and "delocalize" over a vast range of subjects, yet at some point his mind would seize on a particular point and, like the wave function, collapse inward and bring great intensity and clarity to a specific issue.[18] Basil Hiley preferred the image of a helix. Bohm, he told me, would enter very deeply into a particular topic for around six months, until one day he would be found at work on something quite different. In time the original idea would return, and Bohm would go into it again at great depth. Finally, after a number of cycles along the helix, Bohm and Hiley would crystallize their idea, and their scientific paper would be written very quickly.

Writing *Science, Order and Creativity* together enabled us to look more closely at the whole question of order. The book made a distinction between descriptive and constitutive orders. In the case of a house, the descriptive order is the architect's plan, while its constitutive order is the arrangement of bricks, windows, pipes, and wires. Bohm's ideas on mind and matter—soma-significance and signa-somatics—made use of the duality between descriptive and constitutive orders. What appears, at one level, as a shift in the content of consciousness is, at another, an objective change in its structure. Bohm's maxim "a change of meaning is a change of being" also expresses the duality of subjectivity and objectivity, description and constitution. Of even greater significance was the idea of a generative or creative order. This was an avenue we had hoped to pursue over the years. It was certainly in our minds when we discussed writing a second book together.

As with *Science, Order and Creativity,* we first of all met and talked—this time at the Bailey Farms Institute in Ossining, north of New York City. We tossed around a variety of ideas as to what exactly would be the theme of the book. Suddenly its content appeared entire to both of us. It happened very rapidly, as we were walking along the driveway from the house to the road. The book would be called *The Order Between,* and it would deal with the unknown creative ground between dualities.

Consciousness divides the world into oppositions, an idea that was familiar to Bohm from Hegel. But dualities are traps of the mind, nonnegotiable positions between individuals and nations, and blocks to creativity. One resolution is to seek compromise in some middle ground. What Bohm and I now sought was an order that lies both beyond and between. We rapidly wrote down a list of dualities we would discuss in the book:

Nominalism and realism in the Middle Ages

Romanticism and classicism

Shakespeare's exploration of extremes in *Romeo and Juliet*

Coleridge's discussion of fantasy and imagination

Reason and imagination

Music as an exploration of orders between

Environment versus heredity in education

Free choice versus no choice

Holism versus reductionism

Absolutism versus relativism

Mind and body

The mental and the material

The conscious and the unconscious

Jung and Pauli's discussion of an order between physics and psychology, and Pauli's image of the *speculum* that lies in the order between two worlds

Deconstruction and Structuralism

Buddhism and the exploration of a middle way

Darwinism and Lamarckism

A plan of the book was submitted to our previous publisher, Bantam Books, who expressed interest. But then Bohm grew uneasy about being bound to a contract and about accepting money in advance of publication. What if the book were never completed? he asked me. What if it were no good? Would he be forced to pay back the money? Bohm was naturally given to worry but this time his anxiety seemed almost irrational. It was as if he anticipated the darkness ahead that would in fact prevent us completing the work.

In this same period Bohm and his collaborators were looking for a way to develop his ontological interpretation. Bohm now spoke of the quantum potential in terms of a field of "active information." The quantum potential contains information about the entire experimental situation in which an electron finds itself. When the electron moves along its path, it is responding to this pool of information. Bohm compared the situation to an ocean liner approaching a port. The vast output of its engines drives the ship through the sea, but its specific course is determined by radar signals from the harbor. The energy associated with these signals is negligible compared with the power of the engines. Yet radar signals are rich in information and dictate the ship's direction. In this sense a weak but highly informed energy is able to give form to a gross "unformed" energy. Something analogous occurs, Bohm suggested, with the motion of an electron.

But in this example information is not exactly "active." The point Bohm was making is that before a quantum measurement is registered, there are vast possibilities for alternative outcomes, each of which is present within the field of information associated with the quantum potential—each of them is potentially "active." But after the measurement has been registered, only one of the possibilities becomes an actuality. Information about alternative possibilities is still present within the quantum field, but it has ceased to be in an "active form" and cannot affect the future of the quantum system.

That information and its activity have an objective nature is a powerful new idea. Previously, information had been used in physics in ways that were not always entirely clear. In the discussion of entropy, for example, physicists equated an increase in entropy with a diminution in the order of a system. But is there a clear, objective way of defining such order? As with harmony in music, one person's notion of order may be another's disorder. Does information have a real, objective existence in the physical world, or is it merely a feature of the physicist's description of nature?

Bohm was now suggesting that information can in fact be truly objective and that it occupies a central place in physics. Information must be placed alongside energy and matter as one of those factors underlying the processes of the universe.

In Bohm's theory the electron is able to "read" and respond to this information. But this means that it is no longer an elementary, structureless "billiard ball" but has considerable inner complexity. Bohm liked to make a somewhat ironical aside to astounded scientists that the electron is at least as complicated as a television set.

Bohm also applied this idea of active information to the human immune system and the body's ability to maintain its coherence. Active information also plays a significant role in human society, he believed, citing environmental pollution to clarify what he meant. Rather than merely cleaning up a river, one should eliminate the source of the pollution upstream. Likewise, he thought, one should focus not so much on solving the particular problems of human society as on getting rid of their generative source—which he called "pollution," or misinformation. Just as the human body uses an active pool of information (the immune system) to clear up the misinformation that generates sickness, so a field of active information can be used to heal the problems of society. Bohm believed that dialogue process was the active information that would clear up society's "pollution" at its source.

In the mid-1980s Bohm was actively pursuing a number of different ideas and appeared to have recovered from his breakdown. He was still in contact with Krishnamurti, even if their relationship was not as before. In 1985 he bade farewell to the Indian teacher, who was about to make his annual trip to India and Ojai. While he was speaking in India, Krishnamurti noticeably lost weight, and on January 10 of the following year, he flew to Ojai, where his illness was diagnosed as cancer of the pancreas. By February 17 he was dead, leaving Bohm with strong unresolved feelings about his former friend.

Since Bohm had once been a regular visitor at Ojai, a "Bohm committee" now persuaded him to return for six weeks out of each year. To give a formal structure to his visits, they were advertised as a series of seminars. This label made Bohm nervous. "I've had no new ideas, and people have paid to see me," he fretted.[19]

In the end things went perfectly well, and Ojai became another laboratory in which he could research dialogue and develop new ideas on psychology and the ego process. His work was developing so rapidly that he planned a book on the nature of the self and its relationship to thought. Unfortunately he never wrote it; all that remains of his work in this area are transcripts of his many seminars and dialogues, some of which are being published posthumously. They are, however, only a dim reflection of what Bohm might have achieved in this new field, had he still been at the height of his powers.

The self, the supposedly objective subject, is an illusion created by the speed and hypnotic effect of thinking, Bohm argued. If the thinking process could be slowed down, something other than the observer/observed duality would come into existence and give attention to consciousness. Bohm interpreted this nondual state as a sort of seeing from the inside.

To Lee Nichol, Bohm waxed enthusiastic about the way his ideas were developing, yet to David Moody he lamented that he was unable to see the whole picture and transform it into a book. Bohm's missing element, Moody believed, was still the nature of psychological time—the high speed at which conditioning operates on thinking, and how this process could be slowed down.

While Bohm indicated to both Nichol and myself that a transformation had taken place in his own consciousness,[20] he told Moody that nothing radical had taken place. Moody felt that while Bohm had profound verbal insights, he was not always able to perceive their truth directly.[21] Joe Zorskie once told Bohm that he himself felt separate from the rest of the universe and that weaving a web of words around the world was very different from directly perceiving its unity. Bohm nervously admitted that he felt in the same boat.[22] On this score he appears to have presented himself in different ways to different people.

In the wake of Krishnamurti's death, those who had been close to the Indian teacher expressed considerable interest in each other's consciousness. Was it possible that someone had experienced a "transformation of consciousness"? Shortly before his death the Indian teacher had declared that no one had ever truly understood his teaching; no one besides himself had experienced transformation. There could be no compromise, no partial or provisional transformation. If it occurred at all, it was immediate and total.

Bohm, for his part, asked himself why transformation had happened to Krishnamurti alone. Who or what had the Indian teacher been? Had Krishnamurti constantly lived in the transformed state of total mental stillness in which the "intelligence" operates? (Krishnamurti once told the neuroscientist Karl Pribram that, during sleep, his brain was totally silent; its cells did not function, and he never dreamed.[23]) Or had he slipped in and out of this transformed state, sometimes trapped by conditioning, yet able to sustain an intensity far beyond that available to ordinary people? It was variously said that Krishnamurti was an entirely different order of being, or that he was an empty vehicle used by some transcendent intelligence. To Mary Cadogan, Krishnamurti was simply human, albeit an exceptional teacher of extraordinary intensity, intelligence, and sensitivity.

Bohm believed that people like Newton and Einstein did experience a transformation of consciousness, yet they rapidly fell back into conditioning again. Why, he wondered, had such conditioning not "stuck" with Krishnamurti? Was it something to do with the many childhood illnesses that had left the boy mentally dull? Or had he found the Theosophical teachings to which he had been exposed so absurd that they had no chance of influencing his character as "conditioning"? But then, had Krishnamurti really been free from conditioning? At times he had seemed very much to be trapped by his ego.[24]

Such questions burned for Bohm. If Krishnamurti's mind had not been fundamentally different from everyone else's, then he, Bohm, had been hoodwinked and was guilty of naïveté. If Krishnamurti's mind were ordinary, the whole validity of his teaching was called into question. Bohm turned the issue over and over in his mind and finally concluded that he could find no holes in the teachings and no evidence that Krishnamurti was not what he had claimed to be.

Nonetheless, another nagging question remained. Why had no one around Krishnamurti experienced a radical transformation of consciousness? For Bohm, mind is nonlocal; why had the transformation not been passed on? Had none of them been listening properly? Had there been a problem in the way Krishnamurti communicated? Was there something incomplete in the message?[25]

Bohm was now talking with the Dalai Lama, who joked that the scientist had become his physics teacher, although "when the lesson is over I forget everything." His Holiness had a keen interest in modern physics and asked Bohm to explain its key ideas to him. When they met in Switzerland, Bohm developed a high fever, and His Holiness sent for his personal physician, who diagnosed Bohm's blood as too thick. When the doctor indicated that he would send to Dharamsala for medication, the Dalai Lama insisted that the treatment should begin immediately. He took Precious Tablets, wrapped in silk, from a pouch in his room and instructed Saral on how they should be prepared. Bohm found their taste revolting. On another occasion the Dalai Lama sent his personal physician to London, where he examined Bohm at his home in Edgware.[26]

Toward the end of the 1980s, it became clear that Bohm's health was deteriorating. The arteries grafted around his heart were clogging up, and his doctors were talking about angioplasty or drug therapy. Bohm's energy was low, his mood depressed, and he was no longer able to take his habitual long walks. To make matters worse, the English winter and spring had been rainy, windy, and cold, which left him feeling trapped in the house.

There were some high points, however. 1990 saw Bohm elected to The Royal Society. He was also awarded the Elliott Cresson Medal by the Franklin Institute in Philadelphia (the same institute he had visited as a boy) and planned to travel there for the ceremony the following year. Yet these honors were vitiated by his failure to win the Nobel prize. Each year the rumors would circulate that he was short-listed for his work with Aharonov, or for the discovery of nonlocality with John Bell and Alain Aspect. But no call came from Stockholm.

I too sensed the mood change in Bohm. During previous years we had met each October to work on *The Order Between,* but increasingly our collaboration simply did not work out. The more we talked, the further we moved from our original conception. One of Bohm's preoccupations was the Finnish physicist Kalervo V. Laurikainen, who felt that, in his constant desire for order, Bohm did not leave room for what Pauli had called "the irrational in matter." Bohm strongly rejected the idea of "the irrational." It was meaningless, he said—nothing occurred in the universe that was not without order. He talked more and more about the need to discover order, rejecting anything that hinted at chance or randomness.

As the days went on, we circled around the same points without making progress. It was as if Bohm had a personal need to reject any intimation of chance and the unknown. A few years before, we had talked of moving beyond fixed and known forms of order into "an order between and beyond." Now Bohm's creativity seemed to have dried up, and each evening after we parted, I felt totally exhausted. To a psychiatrist the danger signs would have been apparent. "The environment knows" is an old adage; depression spills over onto friends and relatives, who are often the first to express distress.

I voiced my reservations in a letter in August 1989: "We are old friends and I hope that you will speak frankly with me. I feel that our relationship and discussions together are the most important thing and that these will continue whatever happens. If the book still seems a good idea then let's plan for it. But if it now seems inappropriate to you at this time then let us drop it—or do something more modest.

Circumstances and interests may have changed for you and I don't want to feel that we have to go on with something that may have lost its interest or importance for you."*

In March 1991 I received a telephone call telling me that the Bohms were canceling all their travel plans and that I would not see David Bohm in North America that year. He had fallen into the grip of an intense depression.

* Following Bohm's death, Saral told me that he himself had also become concerned about the book, for he believed I had lost interest in it.

CHAPTER 16

The Edge of Something Unknown

FOR BOHM, the grail was wholeness: not that monolithic authoritarian wholeness of universal law and ultimate theory, but a wholeness that was subtle and moving; the universe as an infinity of levels, each qualitatively different yet expressing unity within the multiplicity. Without underlying wholeness the cosmos would make no sense, yet levels cohere and the entire cosmos has an organic quality about it.

Thought fragments wholeness; it breaks apart the material world into elementary particles localized in space and time, and it draws boundaries between nations and creates concepts (boundaries of ideas) within the mind. Fragmentation, for Bohm, was not simply false separation—taking apart what belongs together—but also false unity, or forcing together things that are truly different.

Bohm had begun his search for wholeness in the external physical universe. Yet he had always known that this universe, and all its laws, existed within the microcosm of his own body. His quest lay both without and within. It was a quest guided by illumination, such as had once come to him in the dreams and fantasies of his childhood; it had come again in his seminar at Berkeley and in those intense moments of investigation with Krishnamurti. He sought wholeness in the light, and when that light failed, the effect was devastating. It was a true living death.

A variety of explanations were offered, by friends and colleagues, for the terrible depression that descended on Bohm in the spring of 1991. His health was failing, and he no longer had the energy for creative thinking. Was it a passing phase, he worried, or had his creativity vanished forever?

He was deeply concerned about events in the Persian Gulf. Saddam Hussein had invaded Kuwait: an unpredictable madman, he had to be stopped by force. Maybe biological weapons would be unleashed, maybe even on Israel. Would the multinational forces be successful in their action, or would it all end in indecisive bloodshed? Several of Bohm's friends were shocked to learn that he supported the U.S. military intervention. Their hostile reactions struck Bohm as almost another form of betrayal.[1]

As events in the gulf unfolded, Bohm's distress was exacerbated by the television news coverage. Concerned for the safety of her relatives in Israel, Saral hoped to recognize on TV the areas on which SCUD missiles were falling. While Bohm believed that the United States was acting correctly, he was disturbed that war had been necessary and, when the bloodshed was over, that Hussein was still in power.

Bohm genuinely believed that dialogue groups could make a difference. They would help resolve conflicts by moving directly to their roots and act, in a nonlocal way, on human society. Now, however, in the light of global events, his approach seemed futile, and he questioned the value of this work.

Those of his friends with an amateur interest in psychotherapy wondered if the depression meant that Bohm was confronting the actuality of his life and work. Perhaps he felt he had not achieved the potential of his childhood dreams and visions. He was in the grip of a major spiritual crisis, they believed, yet his very illness suggested how his life and work could evolve further.* Tragically, acting on such insight is exactly what a depressed person is paralyzed from doing. From within a profound depression, it is impossible to make choices

* In contrast, Hiley believed that what could be termed a "spiritual crisis" did not fully emerge until Bohm was discharged from the hospital. It was then that he began to examine and question his life's work.

or to be clear about alternatives. The disease of depression is so serious that, paradoxically, a depressed person who feels that death would be preferable may not even be able to summon up the necessary will for suicide until they actually recover somewhat.

Another, more mundane explanation is possible. Often depression is a cyclic disorder, and its onset is connected with metabolic processes in the body. To be sure, some depression is reactive, a response to a disastrous crisis in a person's life. It can also be a consequence of the isolation and the loss of faculties of the very old. But in so many cases its onset is determined by the body itself rather than by the onslaught of external events. As blackness descends, crises and disappointments are so magnified as to appear insoluble. In some cases the world outside seems to crush and oppress. In others a person is so incapacitated as to be indifferent even to physical pain. But it must be remembered that Bohm's physical health had been declining. As they worked together, Hiley could not help noticing how weak and grey-faced his colleague had become.

As Bohm's depression grew in severity, he experienced additional shocks. *Lives in the Shadow with J. Krishnamurti*[2] appeared, claiming that for more than twenty years the Indian teacher had kept a mistress. Rajagopal had been the manager of Krishnamurti's financial affairs, and in 1932 his wife, Rosalind, and Krishnamurti became lovers. Rosalind became pregnant on several occasions and suffered miscarriages and abortions.

Bohm refused to read the book, but plenty of friends were only too willing to telephone him with the spiciest tidbits. The effect on Bohm was devastating. He was particularly distressed by the abortions and wondered if Krishnamurti had talked Rosalind into having them.

Krishnamurti had been hypocritical, Bohm felt. Any physical relationship with a woman would have been contrary to his public statements—after all, he had spoken to Bohm of the importance of celibacy. Bohm had placed Krishnamurti on a pedestal and treated him as a father figure, when all along the man had been an impostor. Not only had Krishnamurti been less than ideal, but—Bohm now felt—he himself had seen things incorrectly and was guilty of falseness.[3] (It is important to note that Bohm expressed these opinions during the agony of his depression. They do not reflect his final and more reflective opinion of Jiddu Krishnamurti.)

The only thing Bohm could still hang on to was physics. Basil Hiley believed that Bohm had already had an intimation of his death and knew that a definitive statement of the ontological interpretation would be his final work. The book he had written with Hiley was to be the summation of work that dated back to 1952. It would rebut decades of criticism and justify his original position.

The ontological interpretation offers an account of the actuality of microphysical events, something the Copenhagen interpretation denies. An important application is the quantum measurement problem—how is it that something definite comes out of probabilistic potentialities? This problem had been solved by Bohm's approach in the 1970s. In Bohm's theory active information, associated with the quantum potential, guides the quantum system into a definite outcome. When this new state is registered in the macroscopic world, the information becomes inactive. Potentiality is transformed into actuality without any need for human observers.

Their book together was virtually complete. But now, in his state of extreme mental anxiety, Bohm worried about what is known as the classical limit. He had always stressed that quantum theory connects to our early ways of perceiving the world and that it is appropriate to new orders of description in physics. Yet another important point must be made: Any theory of the microscopic world must also contain, within its limits, a description of our own classical world made out of rigid objects located in space. Bohm himself had stressed the importance of the implicate order, but he also pointed out that an explicate order unfolded from it. So too the world of classical physics must emerge out of Bohm's ontological interpretation. Large-scale objects had to be well defined and localized over long time periods. Active information appeared to resolve that problem—when active information becomes inactive, then well-defined outcomes remain. But did information really remain inactive over an infinitely long time period?

It was only as Bohm and Hiley began to think about the origin of the universe within their causal interpretation that this problem began to resurface. Was it possible, they began to wonder, that under certain conditions a degree of ambiguity enters the quantum system? There was no difficulty in demonstrating that, within Bohm's theory, a classical limit exists in our world. The problem arose during the first instants of the creation of the universe. Under what conditions could they ensure that definite events would occur without a residual interference from inactive information? Put another way, how could they bring about a collapse of the quantum wave function before large-scale bodies existed? (This problem of the initial state of the cosmos presents outstanding difficulties in conventional quantum theory.)

Bohm and Hiley tried out a number of different scenarios for this initial state and it was clear that in order to solve the problem they had to make a number of assumptions. As far as Hiley was concerned, they had resolved the issue while, at the same time, exposing a particular deep problem about the way in which our universe was born. But, as his depression grew, Bohm became obsessed with this point, blowing it up out of all proportion until it loomed in his mind as the major defect in his whole work. For decades he had criticized Bohr, claiming that the Copenhagen interpretation of the measurement problem was obscure. But now he felt guilty of the same crime. The book must not be published. He told me he had deceived colleagues and students. Everyone's reputation would be ruined.

Basil Hiley understood the problem but did not feel that it was insurmountable. It was simply a matter of getting down to more work. But Bohm knew that his energy had gone. He could no longer rise to the challenge. His life had collapsed around him.[4]

Now Bohm took to telephoning friends and telling them his work was worthless. To me, he explained his scientific anguish using a remarkable symbolic image. In grappling with the question of whether inactive information could still have an effect when extrapolated to infinitely long times, Hiley had convinced him that, where systems in interaction are concerned, there is always a definite, unambiguous, and objective outcome. But what, Bohm asked, of systems not in interaction with each other, such as distant stars? The motion of each individual atom in a star is guided by the active information associated with the quantum potential. Attracted to the example of a star, Bohm now argued that, in an astronomically long time scale, the individual particles within a star fluctuate, and so does the center of mass of the star itself. In other words, unless they were fixed by observation, the stars themselves would move, fluctuating around their positions in the sky.*

It was a truly staggering symbol for Bohm to have conceived. Essentially he was saying that he could no longer hold the stars steady in the sky. His lifelong search had been for order. Now his own theory was generating chaos in the heart of the universe. Increasingly this failure became the leitmotif of his conversations with friends, colleagues, and psychiatrists. When people encouraged him to press on, Bohm replied that he was too old and that his health and energy were such that he would never be able to work creatively again.

Hiley continued to revise the book, feeling that no major problems remained. Bohm's energy was still low but he continued to make useful suggestions during these discussions. While there were deep and interesting questions in the book, Hiley believed that what had hitherto been implicitly hidden within quantum theory was now out in the open.

During the initial period of his depression, Bohm was in regular psychoanalysis with Patrick de Mare. The sessions were harrowing. On the occasions Saral went with him and sat in the waiting room, she heard her husband weeping and even crying out like a child in pain. His agitation became so intense that he was unable to sleep at night, even with sleeping pills, as thoughts raced though his head. Clearly he was unable to travel and fulfill his various engagements. Yet Saral was reluctant to tell people he was suffering from depression. Mental illness, she believed, still carries a stigma, and she was worried that Bohm's critics would use it to discredit his work.[5]

Bohm leaned heavily on David Shainberg at this time, telephoning him in New York several times during the day and for as long as an

* Technically speaking, if the information is not absolutely inactive each star would be surrounded by a diffraction pattern.

hour at night when he was unable to sleep. (Bohm had also called Shainberg many times during his 1984 breakdown.) In the worst of his depression, he could not be left alone for even a minute and would call Shainberg three or four times in the middle of the night. The phone rang, and Shainberg or his wife, Catherine, would lift it to hear, "Hello, I'm feeling terrible." Catherine described the silence that followed, as Bohm waited for her to speak, as one in which he seemed to be sucking out her blood. Finally Bohm would speak about his various anxieties, Krishnamurti, physics, Kuwait, and the violence in the world. As he talked, it was as if the world itself were coming to an end. In the midst of his pain, he was still questioning. What had been the meaning of his life? Had he really understood Krishnamurti's teachings? Had he given up his authority to Krishnamurti?[6] The situation reached a crisis point, and Saral received a telephone call from de Mare asking her to bring Bohm's nightclothes to his office. Her husband was suicidal, and must be admitted into the Maudsley Hospital. When she received the call, Saral burst into tears but managed to compose herself by the time she arrived at de Mare's office. She asked him if hospitalization was really necessary—she could continue to look after him at home, she argued. De Mare was firm that Bohm required constant attention.

The institution he entered is the Section of Old Age Psychiatry run by Raymond Levy, housed at the Institute of Psychiatry associated with the Maudsley Hospital and the Bethlem Royal Hospital. The Bethlem, founded in 1247, was the original "Bedlam," the first mental hospital in Europe. Bohm was admitted on May 10, 1991, and remained in their care until his final discharge on August 29 of that same year.[7]

By the time the Bohms arrived at the Maudsley, it was late in the day and the duty doctor could not be found. Saral herself was in a distressed state, and the ward, located in a chalet containing some ten or twelve patients, each with their own room, seemed old. Bohm protested that he did not want to stay and asked to be taken home. But Saral persuaded him to remain, at least until the doctor arrived, while the head nurse, exaggerating in a friendly way, said that he could lose his job if Bohm left. Two of the women patients also urged him to stay. Eventually the duty doctor arrived and told Saral that her husband was "a very sick man." As she drove home, she felt terrible at having to leave him there.[8]

In his case notes on Bohm, Professor Levy, the supervising physician, recorded his earlier periods of depression. One at the age of twenty-six (in 1943) had lasted for two years, during which period he had undergone psychotherapy in San Francisco. Another, in 1956 at the age of thirty-nine, following his marriage, had required psychoanalysis for six months. After his last breakdown he had been de Mare's patient for six or seven months. Levy believed that Bohm's mother had been misdiagnosed as schizophrenic and had, in fact, been suffering from manic depression, a disorder with a genetic component. He also noted that Bohm connected his depression to the Gulf War and the attacks of SCUD missiles on Israel.[9]

During therapy Levy observed Bohm's extreme need for a guru to lead him, and having learned of his history of projections on to male figures, including earlier therapists, he refused to accept this role. His conclusion was that Bohm's illness called not for a probing analysis but for protection in a hospital atmosphere and the administration of antidepressants.[10]*

Bohm's initial medication consisted of the antidepressant sertraline at 100 milligrams per day. Many of the more powerful modern antidepressants are able to ameliorate various forms of serious depression, sometimes within a matter of days. Bohm was not so fortunate since some of these drugs have side-effects on the conductivity of the heart and were contraindicated in his case. Of those that could have been used, some produced uncontrolled trembling, while others had only short-lived beneficial results.[12]

Saral visited David in the hospital every day, and he remained in constant telephone communication with his many friends, such as David Shainberg and Paavo Pylkkanen, a former student at Brockwood Park who was now studying the philosophical implications of Bohm's thought. When Don Factor visited the hospital, Bohm told him that he had done a terrible thing, a horrible thing, for

* Bohm's friend Booth Harris, who often talked to Bohm about psychoanalysis, believed that while Bohm had worked hard at psychoanalysis, his mind was "a trap" for an analyst. Bohm could quickly develop a sense of camaraderie with a therapist, thereby subverting the whole process.[11]

which he could never be forgiven. At the time he would not tell Factor what it was, but later he said that his ontological interpretation was flawed, thirty years of work were wasted, his students deceived, and careers destroyed.[13]

Joan Evans, another frequent visitor, saw Bohm's life and work from the context of the psychosynthesis movement. She felt that while Bohm had spent his intellectual life at an edge, moving into those terrains of physics where others would not venture, he nonetheless retained a great fear of the unknown. She saw his depression as a spiritual crisis, the result of his encounter with uncertainty, that final step he felt unable to take.[14]

Drug therapy was not as successful as they had hoped, but at some periods, for a short time at least, Bohm improved to the point where he was allowed home for the weekend. On one of these occasions Lee Nichol flew from Ojai to London.

The moment he saw Bohm, Nichol was taken aback by his pale and weakened condition. Yet as Bohm began to talk, he revived and his energy increased. Nichol almost believed that he was returning to his old self—until suddenly the energy became transformed into agitation and Bohm began to pace the room and talk in a loud voice. He was obsessed with money; taxes had to be paid in the United States. This topic became mixed up with others to such an extent that his thinking no longer appeared logical. His talk was a collage of Krishnamurti, the Gulf War, and the ontological interpretation. Each time Nichol tried to reason with him, Bohm became more irritated. He raised his arms, scratching himself in agitation and saying, "Now I'm vulnerable—they're going to be able to get me." He talked of the intelligence agencies of the U.S. government. His work in physics was disintegrating, and his association with Krishnamurti would be exposed. Everything around him was falling to pieces, and "now that I'm weak . . . they've been waiting. Now they have their chance to get me."

Bohm's thinking was so out of control that he was suffering badly. Neither Nichol nor Saral could calm him, and Bohm told his wife to phone for a taxi since he wanted Nichol to leave.

On the following day they met again. As they ate together, Bohm appeared calmer, even speaking on the telephone to his friend Georg

Wikman in Sweden. Yet Nichol noticed that Bohm appeared startled each time the telephone rang—and the phone rang frequently. Soon the agitation began again, with Bohm telling Nichol he was under constant surveillance and that people were watching him and bugging his telephone calls. This time his thinking remained more rational. He spoke of what had happened to him during and after his appearance before the House Un-American Activities Committee, and of the deep mark it had left upon him. It is easy for people to be set up and framed, he explained. Nichol wondered if Bohm felt guilty about some past political action he had performed, but soon it became clear that the source of his distress was the vivid memory of that atmosphere under which he and his friends had once lived.[15]

As has been mentioned, Bohm could not be prescribed the most powerful of antidepressants. In spite of the medication he was taking and the hospital protection, his obsessional ideas increased in intensity. He stopped eating and drinking and was increasingly suicidal. It was clear to Levy that more radical treatment was needed. Her husband was not improving, the doctors pointed out to Saral, and they wanted to try a course of ECT (electroconvulsive therapy or "shock treatment"), to be followed by cognitive therapy. Bohm was not sufficiently well to make the decision, so they were relying upon Saral. She phoned her friends for advice and in particular relied on the judgment of David Shainberg.[16] At first Shainberg objected to the treatment, but after talking to his colleagues, he agreed that, under the circumstances of Bohm's admission into the Maudsley and in view of his age, ECT was probably his only option.[17]

In anticipation of the treatment, Bohm became even more agitated. He was frightened that he would experience memory loss and be unable to continue with his creative work. I spoke to him a day or so before his first treatment and a few days afterward. The actual experience turned out not to be as bad as he expected.

Levy prescribed a course of fourteen bilateral ECTs, and as far as he and the staff were concerned, Bohm showed a marked improvement. Referring to his case notes, Levy wrote "at the end of the course of ECT his sleep, appetite, energy and mood were essentially normal and his previously held abnormal beliefs and suicidal ideation had resolved." Levy did, however, agree that Bohm had not been able to come to terms with "the fact of his age and with the drying up of his creativity."[18]*

Discharged from the hospital on a daily 50-milligram dose of sertraline, Bohm worked with a cognitive therapist who gave him a series of mental exercises to perform first thing in the morning. The purpose was to control the way he was looking at the world. Bohm enjoyed working with her and even explained his own theories to her.[19]

Clinically speaking, the major effects of Bohm's depression were cured, but the memory defects he suffered over the next months troubled him greatly. On one occasion a German television company was filming an interview with him at Birkbeck, but after only a few minutes Bohm left the room in distress, unable to continue. When Israeli television visited, Bohm appeared so frail that Hiley sat in on the interview to give him confidence.[20] In the spring of the following year, during a group discussion, Bohm turned to me and whispered, "What did I used to think about this?"

As far as Hiley was concerned, Bohm's major problem now was not so much the effects of the shock treatment or his residual depression but the lack of energy he needed to continue his work. Not many people realize just how much energy is required to pursue a scientific or philosophical idea in a totally focused way for several hours each day. Hiley and the students were making advances on the very work that Bohm had initiated, yet he simply could not keep pace with them. His energy had disappeared, and he was not even able to follow through his own ideas. Sitting in on a discussion, he was overwhelmed and lost. Later he asked Hiley, "How can I keep up with it?"[21]

Where once he had radiated ideas that inspired others, Bohm now sat in Hiley's office and asked, "Basil, do you think I'll ever get better?" Hiley realized Bohm's health was so delicate that he would never again be able to do creative physics as he had done before. It was particularly ironic that this should have happened at a time when much of their work was bearing fruit. After his friend's death, Hiley

* In Saral's opinion he responded well at first, he then reached a plateau with no further improvement.

lamented, "If only Dave were here, he could have cleared this up in an hour's discussion. Now it's going to take us months."[22]

While Bohm exhibited a general sadness at the loss of his abilities, Saral noticed positive changes in his attitude toward life.[23] Bohm had never been a harsh or assertive man, but now he was even more gentle and tender. Where once he had spoken in abstractions and generalities, it was becoming easier for him to talk about his feelings and express his love for her.

His physical health was weak, but the spring of 1992 found him relaxing in the Ojai sun and talking to his friends. Lee Nichol spent a great deal of time with him, encouraging him to walk a little each day. While Bohm was not his old self, he continued to probe ideas on psychology in dialogues and discussions.[24]

From Ojai, Bohm moved on to the Fetzer Institute in Kalamazoo, Michigan. There he told me that the shadow of his depression still lingered. The last year had been a terrible time for Saral, and he would never subject her to that again. If the illness returned, he would kill himself.

He was worried about *The Order Between*, the book we were supposed to complete, but I persuaded him that this time, at least, we would simply meet as old friends. And so we gossiped and talked of old times. At occasional moments his intensity returned and he talked with enthusiasm about generating mathematical structures in space-time and using them to derive the Lorentz transformations—which are the basis of special relativity. Although he was clearly weak, on those occasions his mind could still run rings around mine.

While we were talking together one afternoon, Saral summoned him to the phone. She whispered to me that it was the Technion in Israel, awarding Bohm an honorary doctorate. His conversation was not unlike that of the previous day, when he had been offered a complimentary copy of Kalamazoo's daily newspaper.

> Hello. Yes, this is David Bohm speaking.
>
> Yes? Oh, I see. Yes, well thank you. (Pause)
>
> What do I have to do? (Pause)

Oh, I have to go there? Is that really necessary? Oh, very well.

Good-bye, then. Erm, it was very kind of you to offer—erm—what you offered. Well, good-bye.

He immediately returned to the topic of our conversation. I imagine the party at the other end was not at all certain if Bohm had accepted the honor or not!

The timing of Bohm's visit to Kalamazoo was fortunate, for it was also the occasion of the first of a series of circles between Native Americans and Western scientists. Leroy Little Bear, a Blackfoot philosopher, and I had organized the circle, which arose out of our earlier discussions on the philosophy and physics of the Blackfoot people. When the Native participants learned of Bohm's presence, they made him guest of honor and treated him as a distinguished elder. One of them, distressed at the condition of his heart, indicated that during the opening ceremony, when sweetgrass would be passed around the circle, we should sacrifice part of our energy so that it could be given to Bohm.

The Native participants questioned Bohm about the ethical and moral dimensions of Western science. As one of them, Sa'ke'je Henderson, argued, traditional Native ideas of harmony and balance indicate that if order is created in the laboratory, then disorder must be created somewhere else. How, he wondered, does science take responsibility for this? There was also concern at the way physicists learned about elementary matter—by colliding particles together in a particularly violent way. We are also communicating with nature, the Elders explained, and entering into alliances with her energies. What are the moral dimensions of such actions?

As the discussions continued, Bohm learned about the process-based worldview of the Blackfoot, Ojibwa, and Micmaq. Everything is said to be in flux, and this constant change is reflected in their verb-based language. As Sa'ke'je put it, "I can spend the whole day without ever uttering a noun." At last Bohm had found a people whose metaphysics strongly mirrored his own, and whose language, not to speak of the role it played in their reality, echoed his own rheomode. Both Bohm and Saral were moved by the deeply spiritual outlook of the participants. Speaking about the arrival of the first Europeans, Bohm remarked, "It would have been better if we had never come."

On the final day of the meeting, gifts were exchanged and a drum was set up beside a lake. As the Potawatomi people sang and drummed, several of the participants danced, and to my surprise I saw that David Bohm had joined the circle. Dressed in his tweed jacket, he was dancing to the beat of the drum—the drum that is both the human heart and the beat of the universe. Over the weeks that followed, Bohm talked about the deep impression this meeting had made on him.*

Paul Grof, a psychiatrist who specializes in bipolar depression, was present at the Fetzer Institute for the same meeting, and Bohm asked him to conduct tests and assess his condition. Grof concluded that while the acute stage of his depression had lifted, Bohm's depressed feelings were still present. Bohm's level of medication, he thought, was delaying his recovery.† It was also around April, when Bohm's medication was changed, that Hiley noticed an improvement in his general attitude. Where before he had been extremely negative, now he returned to that chapter on the classical limit that had so troubled him in the hospital. He began to make creative suggestions that he and Hiley revised together. While he was never to regain his full powers at least his interest in physics had been stimulated.

Therese Schroder-Sheker also met with Bohm. She is a professional harpist who studied the therapeutic use of music in medieval medicine and practiced it herself. Therese had first introduced this practice into a Denver hospital and was now supervising a Department of thanatology in a hospital in Missoula, Montana. One evening she sang and played to Bohm until he slept. Based on her long experience of the dying, Therese concluded that he was a liminal personality—a threshold dweller who was dipping in and out of his life on earth and was already entering a world beyond.[26]

* Bohm died a few weeks before the second of these circles, held in Banff, Alberta. When the participants sat together on the first evening, it was discovered that, by chance, one chair had been left vacant. This was Bohm's chair, and several of the speakers addressed him directly as if his spirit were present.

† Raymond Levy did not agree with Dr. Grof's diagnosis. He believed that while Bohm may have been experiencing a temporary relapse, he was recovering according to expectations.[25]

But this picture of Bohm's condition needs some qualification. As we sat together—at a restaurant, perhaps—he still made his familiar witty, dry comments. If the talk turned to Krishnamurti, he grew preoccupied with his questions but was still able to discuss the Indian teacher calmly. Yet when guests joined us at dinner, Bohm fell uncharacteristically silent and made few comments, even when a topic was introduced specifically to interest him. On those occasions he had the appearance of a man worn out by life, unable to summon the interest to speak.

Bohm's next trip was to Israel, where he received his honorary doctorate, and then Prague to speak at an international meeting of the Transpersonal Association.* Bohm, particularly nervous about his talk, found the weather hot and humid. Over the next few days he became physically distressed and developed a cough. It looked as if he had picked up a virus. Back in London, he was diagnosed as having suffered a heart attack. Admitted to Edgware General Hospital, he was placed on an intravenous antibiotic drip for viral pneumonia. Saral was warned that his heart was in such a serious condition that he could die at any time. From now on his blood had to be checked regularly, and he was placed on the blood thinner Warfarin.[28]

Back at home again, Bohm was too ill to travel to Birkbeck. He was visited by Paavo Pylkkanen, who discussed the theory of cognition they had been attempting to develop.[29] Hiley was another regular face around the house. By now Bohm was convinced that the problem of obtaining a classical world out of his theory had been resolved. Together they put the finishing touches to their book, *The Undivided Universe*.

Making final corrections was a frustrating business for Hiley, since Bohm held in complete disregard the academic convention that all possible references must be given to those who have done related work in an area. A highly original thinker, Bohm spent little time reading scientific papers, reviews, and the many manuscripts that

* Bohm's later writings and *Wholeness and the Implicate Order* had been prohibited under the Communist regime, but they were studied by a group of dissidents, including Václav Havel and his brother, Ivan, a mathematician who later established a group at Charles University in Prague to study the relationship between consciousness and matter.[27]

were sent to him. Most of his inspiration came from informal discussions and his own thinking. When he did read a paper, it might spark off in him some idea that its authors never intended. On other occasions he would simply absorb and metabolize a concept, making it his own. The task of tracking down all the references to previous work had fallen to Hiley.*

By October, Bohm's health had improved a little, and he could take short walks with Saral. He told Pylkkanen that he was finally happy with his ontological interpretation. Now that it was completed, he said, they should expand their planned paper on cognition into an entire book. He also began to watch videotapes of his and David Shainberg's discussions with Krishnamurti.

Krishnamurti had spoken of the ending of thought, and of the mind that becomes silent so that something qualitatively new can operate to transform consciousness and mutate the brain. Over the years Bohm had accepted that this process can and does happen. Now he was asking "What is it that sees in this nondualistic state? What is it that observes consciousness?" In a telephone call to Lee Nichol, he speculated that in the end, this new mode of consciousness might be nothing more than just another variation of thought. This remark seemed to fly in the face of everything Bohm had ever said, and Nichol wondered if, at the end of his life, Bohm simply wanted a final explanation, one that left no room for uncertainty and allowed a closure of his work.[30]

Bohm also telephoned Don Factor. During his depression he had sometimes called the whole dialogue process into question. Now he felt that dialogue had to confront the fact that our notion of the self was wrong. There was indeed a self, but this self is not an object but an entire mental process, an ongoing activity. Insight, he repeated, brings about a radical change in the brain, not simply at the level of neural pathways but, he told Factor, right at "the quantum level."

As Bohm spoke, Factor sensed that he was seeking a transcendental breakthrough. Dialogue by itself was no longer enough. There had

* Krishnamurti may have supported this attitude. Rather than reading philosophy and quoting other people, Krishnamurti believed that one should go into an issue directly. One should not allow it to be filtered through the theories and opinions of others.

to be something beyond even dialogue. One tries everything, he told Factor, and nothing seems to work; then a level of frustration arises that is so great that it breaks down the entire system. It is at this point that a radical change of meaning can occur. Such a change can never be brought about by an act of will. It involves opening up to something greater.[31]

On that same evening, October 26, 1992, Bohm watched the videotape of the last of the seven dialogues he and Shainberg had made with Krishnamurti at Brockwood Park over sixteen years before. The subject was death and what may lie beyond. Afterward he turned to Saral and said, "We should have gone on talking."[32]

The following morning Saral drove her husband to the hospital for his weekly check of Warfarin. He was in the examination room for an unusually long time. When he came out, he told Saral, "They could not get the blood to flow." After lunch he decided to go to Birkbeck to work with Hiley. Saral asked him to phone before he left for home and to take a taxi if he felt tired.

At around six fifteen that evening, Bohm telephoned to say that he felt well enough to use the underground but would take a cab from Edgware station at the other end. His work with Hiley had gone well, and his voice was bubbling with energy. "You know it's tantalizing," he said—"I feel I'm on the edge of something." Around seven thirty the doorbell rang. Saral assumed David had forgotten his keys. Instead, she opened the door to a cab driver who told her, "Your husband has collapsed."

Bohm was lying in the cab, his wallet on the floor. Since it was his habit to take it out as the cab pulled up at his house, he must have collapsed only moments before. She called the emergency services without success. In the end the cab driver took them to Edgware General. Sitting in the back, Saral tried to revive him by talking to him and blowing into his lungs. There was no response except for the fluttering of one eyelid. At the hospital they worked to revive him for half an hour. Then he was pronounced dead.[33]

Bohm had expressed little interest in the arrangements for his funeral, but out of respect for his family, Saral honored the practices of the Jewish religion, which meant a rapid burial. But to have his body released for burial, she needed a death certificate. Bohm had not been in the hospital for the requisite twenty-four hours for the certificate to be signed by one of their doctors, and his cardiologist was away in the United States. In the end the local coroner issued a death certificate. The autopsy report states the causes of death were left ventricular failure, myocardial ischemia, and coronary artery atheroma.

Bohm was buried at the Jewish cemetery in Waltham Abney on a cold but sunny October day. Since Bohm had been a Cohen, a member of the priest tribe, he was given a special corner of the cemetery. Rav Galumb, who conducted the ceremony, had not known Bohm—after all, the physicist had not practiced his religion since childhood. During that particular week, the rabbi explained, a section of the Torah was read that dealt with Noah and the flood. Noah and his family were making a new life, attempting to start something better. From what the rabbi had been told about the man they were burying, he too was concerned with the future of mankind.

The mourners cast earth on the coffin. Saral's sister threw in pebbles she had brought from Israel. Basil Hiley and several of the Birkbeck students helped the family symbolically fill in the grave. The following year a ceremony was held, in which a stone was erected bearing the lines "A great teacher. An inspiration to all who knew him." Bohm's school friend Mort Weiss stood beside Saral at the graveside.

The book Bohm had been revising when he died required only a few more days to complete, and *The Undivided Universe: An Ontological Interpretation of Quantum Theory* was published early the next year.[34] It was well received—a curious irony in view of the way this work had been ignored during his lifetime. Henry Stapp, in a review for the *American Journal of Physics,* wrote, "This book, I believe, will change the way quantum theory is taught."[35] Daniel M. Greenberger, in *Science,* found it to be "a very important book, an attempt to open the minds of physicists to understand that what they are used to is not necessarily true."[36] For Sheldon Goldstein, it was "a brilliant book, of great depth and originality. Clearly written, it provides an unusually incisive account of quantum phenomena. . . . Every physicist and physics student who wants to understand quantum mechanics should read this book."[37] Chris Clarke noted that "the book disturbs the reader, because the profound originality of the thinking differs so much both from mainstream physics and what the new age has made of physics. It could be that it will in the course of time disturb also the progress of science."[38] Already several groups were applying what they termed "Bohmian mechanics" to the quantum theory, and Bohm's ideas were being discussed at international scientific conferences.

Lee Nichol transcribed and edited Bohm's Ojai seminars (from November 31 to December 2, 1990) into a book, *Thought as a System*, which was published by Routledge in 1994.[39] While it is only the shadow of the more comprehensive book that Bohm had planned on the nature of thought and the self, Nichol's introductory essay is valuable for presenting a brief overview of Bohm's thinking at the time.

In the *Thought as a System* seminars (as they are interpreted by Nichol) Bohm discusses the way thought objectifies itself. Repeated thoughts and reactions become "hard-wired" into consciousness in the form of "neurophysiological reflexes." Our human products—science, art, music, architecture, computers, and societies—are the manifestation of "fixed, concrete" thought. The activity of thought, emotion, and intellect, along with the artifacts that thought builds, all form an entire system. Yet a "systemic fault" exists in thought, which "doesn't know it is doing something and then struggles against what it is doing." As a result thought and human society, which is an aspect of thought, are constantly generating incoherence.

The question that Bohm kept returning to in his last years was how to clear up, or heal, the incoherence of thought and how to understand the nature of what he termed "the egoic process." How can thought develop the proprioception of its own activity? He had once accepted Krishnamurti's teaching that it can do so when the mind is silent and something beyond thought comes into operation. What lies outside the system of thought is "insight," "perception," "intelligence," or "active energy." Yet thought's reflexes operate at the social level, where dialogue groups can play a role. Dialogue also opens the possibility of a collective "flow of meaning" within a group that transcends the limits of individual thought.

According to Nichol, "Bohm suggested that the potential for collective intelligence inherent in such groups could lead to a new and creative art form, one which may involve significant numbers of people and beneficially affect the trajectory of our current civilization." Such a suggestion, made toward the end of Bohm's life, vividly evokes his childhood vision of the transformation of society and his delight in contemplating the sweep of human civilizations.

Following Bohm's death, Saral met with the Dalai Lama, who asked her what dreams she had had of her late husband. She told him she had dreamed that her husband was filled with light and that he was communicating with her that all was well. The Dalai Lama appeared pleased, remarking that it was "a good dream." He gave her a shawl to place over a photograph she kept of her husband.

The final words are Bohm's. In 1987 he had written a short piece to be read at the memorial service for Malcolm Sagenkahn, one of his classmates at Penn State. Although their politics were diametrically opposed, they had remained good friends throughout their lives. The piece was also read at Bohm's own memorial service, held at Birkbeck College. Reading the essay was a consolation to Saral.

> In considering the relationship between the finite and the infinite, we are led to observe that the whole field of the finite is inherently limited, in that it has no independent existence. It has the appearance of independent existence, but that appearance is merely the result of an abstraction of our thought. We can see this dependent nature of the finite from the fact that every finite thing is transient.
>
> Our ordinary view holds that the field of the finite is all that there is. But if the finite has no independent existence, it cannot be all that is. We are in this way led to propose that the true ground of all being is the infinite, the unlimited; and that the infinite includes and contains the finite. In this view, the finite, with its transient nature, can only be understood as held suspended, as it were, beyond time and space, within the infinite.
>
> The field of the finite is all that we can see, hear, touch, remember, and describe. This field is basically that which is manifest, or tangible. The essential quality of the infinite, by contrast, is its subtlety, its intangibility. This quality is conveyed in the word *spirit,* whose root meaning is "wind, or breath." This suggests an invisible but pervasive energy, to which the manifest world of the finite responds. This energy, or spirit, infuses all living beings, and without it any organism must fall apart into its constituent elements. That which is truly alive in the living being is this energy of spirit, and this is never born and never dies.

NOTES

WHEN HE WAS IN HIS LATE SIXTIES, David Bohm met with a literary agent in New York who suggested that an intellectual autobiography would be of considerable interest. Bohm was unsure if he wanted to undertake such a task, but he recalled that some years earlier he had audiotaped reminiscences about his life while talking to Maurice Wilkins and myself (David Bohm interviewed by Maurice Wilkins and David Peat, Brockwood Park, England, October 18, 1974; referred to as B/W/P). Maybe speaking into a tape recorder would be the easiest way to begin, he thought, and so in the late 1980s he and his old friend Maurice Wilkins met again so that Bohm could talk about his life. After each session Bohm would think over what he had said and, the next time they met, would try to set the balance straight by modifying earlier remarks. These meetings extended to several hours of audiocassettes (referred to as B/W). As reference material they are invaluable, but it should be added that the picture they give of his life is somewhat clouded by the onset of the depression he was later to suffer. Nevertheless, augmented with his letters, documents, and interviews with his relatives, colleagues, and friends, these cassettes have been an important source for this biography.

Bohm's letters have been useful, not so much for illuminating particular incidents in his life as for revealing the development of his thinking. Unfortunately Bohm rarely kept copies of his correspondence, and it is thanks to the recipients—most notably Charles Biederman, Miriam Yevick, Melba Phillips, and Hanna Loewy—that any of it survives. Often Bohm did not bother to date his letters, so at times it is a matter of conjecture as to when a particular letter was written.

The file on David Bohm presumably compiled by the Federal Bureau of Investigation during and after the war years would have been another important source. During Bohm's time in Brazil and possibly also in Israel and the United Kingdom, Bohm would have been placed under the surveillance of the Central Intelligence Agency. I applied, under the Freedom of Information Act, for all files relating to Bohm from these two agencies. My request to the FBI (#380311) was acknowledged on September 28, 1993, by J. K. O'Brien, Chief of the Freedom of Information–Privacy Acts Section. Despite additional reminders right up to press time of this biography, the file was not made available to me. The response from the CIA was more direct: while not acknowledging that such a file existed, the Agency would not make it available on a number of grounds including that of national security.

During the research and writing of this book I was also able to draw upon the memories of Saral Bohm and David's colleague, Basil Hiley, over many interviews, telephone conversations, and informal chats. Sometimes, in the notes below, a specific interview is mentioned; in others, remarks have been gleaned and amplified over many meetings.

Introduction

1. Interview with Saral Bohm, London, January 15, 1995, together with notes Saral Bohm made on the day following David Bohm's death; interview with Basil Hiley, London, March 1995.
2. Einstein's close friend, Lilly Kahler, related this remark to Saral Bohm. Miriam Yevick also recalled that Einstein's remark, "If anyone can do it [reformulate quantum theory], Bohm can," circulated at Princeton in the early 1950s.
3. "But in 1952 I saw the impossible done," John Bell in *Foundations of Physics 12* (1982) 989–999.

Chapter 1 Childhood: From Fragmentation to Flow

1. Information on the Bohm family history comes from a telephone interview with Irving Bohm, April 23, 1994 and a letter to the author, June 21, 1993. Additional information was given during an interview with Ruth Berman (née Bohm), in Los Angeles, April 1994, and from my many interviews with Saral Bohm.
2. B/W
3. Dr. Mara Sidoli, a Jungian child therapist, points out that such extreme acts of imagination can occur with highly disturbed children. One of her

patients, a very intelligent but disturbed young boy, engaged in fantasy space trips with such intensity that she could not reach him. Only after she persisted in joining him on these imaginary journeys was she able to communicate with him. For Sidoli, the boy's ability to leave Earth, voyage in outer space, and encounter aliens was a form of psychic inflation that compensated for the painful, dislocated conditions of his actual life. Interview with Dr. Sidoli, Assisi, Italy, August 1994.

4. B/W
5. Interview with Mort Weiss, Laguna Beach, California, April 1994.
6. David Bohm, undated draft of a letter to a science fiction magazine.
7. Interview with Mort Weiss.
8. B/W
9. Interview with Ruth Berman.
10. Interview with Edward Bohm, Laguna Beach, California, April 1994.
11. David Bohm left a package of his schoolboy papers with Hanna Loewy when he left the United States for Brazil.
12. Interview with Mort Weiss.
13. B/W
14. A fifty-cent piece was first issued in 1926 to commemorate the Oregon Trail. It depicted a settlers' wagon driving West toward the setting sun. The obverse portrayed a stylized Native American.
15. Telephone interview with Mayer Tope, April 1993.
16. Letter from Sam Savitt, May 26, 1994.
17. B/W
18. Interview with Mort Weiss; telephone interview with Sam Savitt, May 1994.
19. Letter from Sam Savitt and subsequent telephone interview.
20. From David Bohm's schoolboy papers left with Hanna Loewy.
21. *This Fabulous Century: Sixty Years of American Life,* vol. 4 (New York: Time-Life Books, 1969).
22. Francis E. McMahon, assistant professor of philosophy at the University of Notre Dame, expressed his concern about rising anti-Semitism in the United States: "As a Catholic, I am concerned about the spread of anti-Semitism amongst the Coughlinites of my own faith. . . .I cannot fathom the mentality of a Coughlinite who assails the people from which the Founder of his religion came." Frances E. McMahon, "Lindbergh and the Jews," in *Liberty* [Toronto], January 3, 1942.

Chapter 2 Youth: From Penn State to Caltech

1. Interview with Mort Weiss, Laguna Beach, California, April 1994.
2. Ibid.
3. B/W
4. E. T. Whittaker and G. N. Watson, *A Course of Modern Analysis: An Introduction to the General Theory of Infinite Processes and of Analytic Functions* (orig. 1902; New York: Macmillan, 1943).
5. Letter from Greenwall and subsequent telephone interview.
6. B/W
7. Ibid.
8. I have not been able to verify this story independently. Its only source is Bohm's taped reminiscences with Maurice Wilkins.
9. Letter from Greenwall and subsequent telephone interview.
10. Telephone interview with Leon Katz, July 12, 1993.
11. W. R. Smythe, *Static and Dynamic Electricity* (New York and London: McGraw-Hill, 1939).
12. Letter from Melba Phillips, January 29, 1995.
13. Telephone interview with Leon Katz, July 12, 1993.
14. B/W
15. Interview with Basil Hiley, London, March 1994.
16. B/W
17. Telephone interview with Katz.

Chapter 3 A Vision of Light

1. Information on Oppenheimer's life has been taken from a variety of standard biographies, including: *Robert Oppenheimer: Letters and Recollections,* Alice K. Smith and Charles Weiner, eds. (Cambridge, MA: Harvard University Press, 1980); Philip M. Stern and Harold P. Green, *The Oppenheimer Case* (New York: Harper and Row, 1969); Charles Pelham Curtis, *Oppenheimer Case: The Trial of a Security System* (New York: Simon and Schuster, 1955); Martin J. Sherwin, *A World Destroyed: The Atomic Bomb and the Grand Alliance* (New York: Random House, 1975); Peter Goodchild, *J. Robert Oppenheimer: Shatterer of Worlds* (New York: Fromm International, 1985); Joseph Wright Alsop, *We Accuse: The Story of the Miscarriage of American Justice in the Case of J. Robert Oppenheimer* (1954); James W. Kunetka, *Oppenheimer: The Years of Risk* (Englewood Cliffs, New Jersey: Prentice-Hall, 1992). Transcripts of his security hearings can be found in "In the Matter of Robert J. Oppenheimer: Transcript of Hearing before Personnel Security Board,"

Atomic Energy Commission, Washington, D.C., May 27–June 29, 1954 (Cambridge, MA: MIT Press, 1970). This information has been supplemented by discussions with Melba Phillips, Joe Weinberg, Rossi Lomanitz, and my earlier conversations with David Bohm.

2. Telephone interviews with Joe Weinberg, June 26, 1993 and September 11, 1993.
3. Letter from Raymond Levy to the author, March 17, 1994.
4. Telephone interview with Betty Friedan, September 18, 1993.
5. Interview with Miriam Yevick, Leonia, New Jersey, April 30, 1993, supplemented by subsequent interviews and telephone conversations.
6. Telephone interviews with Joe Weinberg.
7. B/W
8. B/W
9. Telephone interviews with Joe Weinberg.
10. Although Bohm accepted Bohr's complementarity at this time, and continued to justify it when he wrote his textbook *Quantum Theory,* he later believed that it had become a block to further developments in the theory. Bohm felt that if Bohr and Heisenberg had not been so quick to resolve the philosophical difficulties of the theory but had stayed with their frustrations, it might have led to a more creative insight.

Chapter 4 From Niels Bohr to Karl Marx

1. B/W
2. Pavel Sudoplatov and Anatoli Sudoplatov, with Jerrold L. and Leona P. Schecter, *Special Tasks: The Memoirs of an Unwanted Witness—A Soviet Spymaster* (Boston: Little, Brown, 1994).
3. Bohm interviewed by Martin Sherwin.
4. Jonathan Rosen writing on Roth in *Vanity Fair,* February 1994.
5. Telephone interviews with Joe Weinberg, June 26, 1993 and September 11, 1993.
6. Aldous Huxley, *After Many a Summer* (London: Chatto and Windus, 1959).
7. Telephone interviews with Joe Weinberg.
8. *Einstein: A Centenary Volume,* A. P. French, ed. (Cambridge, MA: Harvard University Press, 1979) and Abraham Pais, *Subtle Is the Lord: The Science and Life of Albert Einstein* (New York: Oxford University Press, 1982).
9. Interview with Rossi Lomanitz, Albuquerque, New Mexico, April 1994.

10. Manhattan Project Security Records, Record Group 77MED, Box 100, National Archives.
11. Later, Robert Oppenheimer was to tell the security services of a certain "Scientist X" who, he claimed, had passed important information to an agent in contact with the Russians. The security officers eventually identified "Scientist X" as Joseph Weinberg, and after the war Weinberg was arrested. No evidence was offered at his trial, and he was acquitted.
12. Captain Peer DeSilva, QMC, espionage investigation report, August 13, 1943, U.S. Department of War, Military Intelligence Division, DSM Project file.
13. Telephone interviews with Joe Weinberg.
14. B/W
15. Ibid.
16. Interview with Maurice Wilkins, London, March 1994.
17. Ibid.
18. A number of similar anecdotes indicate Oppenheimer's lack of intuition in determining which were the truly important problems and areas of physics.

CHAPTER 5 PRINCETON: A SOCIETY OF ELECTRONS

1. Eugene Gross, in *Quantum Implications: Essays in Honour of David Bohm,* B. J. Hiley and F. David Peat, eds. (London: Routledge and Kegan Paul, 1987).
2. Letter from Daniel Lipkin to the author, April 19, 1994.
3. Ford was later to become the executive director of the American Institute of Physics. Ford telephone interview, 1994.
4. Eugene Gross, *Quantum Implications.*
5. Ibid.
6. Ibid.
7. Gross recalled by Miriam Yevick: interview with Miriam Yevick, Leonia, New Jersey, April 30, 1993.
8. Letter from David Bohm to Miriam Yevick, February 1953.
9. B/W
10. Ford telephone interview, 1994.
11. Lipkin letter.
12. Interview with George Yevick, Ottawa, 1994; and B/W.
13. Interview with Miriam Yevick.
14. Ibid.

15. Ibid.
16. Interview with Roy Britten, Corona Del Mar, California, April 1994.
17. Interview with Christopher Philippidis, Bristol, 1995.
18. Interview with Miriam Yevick. The only detailed account of the affair remains Miriam Yevick's. It is clear that Miriam was deeply and romantically involved with Bohm, but it is not so clear how much Bohm reciprocated her passion. His many letters to Miriam, from his later exile in Brazil, are more concerned with physics and politics than with expressions of tenderness. On the other hand, Bohm was not the sort of person to be demonstrative about his feelings.
19. Interview with Melba Phillips, New York, July 1994.
20. Ibid.
21. Telephone interview with Ford, 1993.
22. David Bohm and Kenneth Ford, letter to *Physical Review,* vol. 79 (1950), p. 745.
23. Interview with Basil Hiley, London, January 18, 1995.
24. It is also interesting to note that just as some of Bohm's ideas in physics arose out of kinesthetic sensations in his body, so too Cézanne spoke of attempting to "realize his sensations" in paint when seated before nature. In both men this act of unfoldment and realization appears to have been accompanied by a degree of inner conflict. From David Peat, "Bohm and Cézanne: The Search for a New Order," lecture given at Edinburgh, March 9, 1995.
25. Sam Schweber, e-mail correspondence, November 1993.
26. Sam Schweber, address made at the Bohm memorial, Brandeis University, December 3, 1992.
27. Interview with Hanna Loewy, New York, 1993.
28. Ibid.
29. Schweber address.
30. In a conversation with Maureen Doolan (New York, April 1994) Hanna Loewy said that she had never really loved David Bohm.
31. Interview with Hanna Loewy.
32. Interview with Miriam Yevick, Ottawa, 1994.
33. Interview with Hanna Loewy.
34. Letter from Bohm to Miriam Yevick, February 15, 1952.

Chapter 6 Un-American Activities

1. Interview with Mary and Rossi Lomanitz, Albuquerque, New Mexico, April 1994.

2. Russel B. Olwell, "Princeton, David Bohm and the Cold War: A Study in McCarthyism," junior paper submitted to the Department of History, Princeton University, May 4, 1990.
3. B/W
4. Interview with Rossi Lomanitz, Albuquerque, New Mexico, April 1994.
5. "In the Matter of Robert J. Oppenheimer: Transcript of Hearing before Personnel Security Board," Atomic Energy Commission, Washington, D.C., May 27–June 29, 1954 (Cambridge, MA: MIT Press, 1970).
6. The names on the list included Frank Malina, Frank Palma, Jacob Dubnoff, Sidney Weinbaum, Orla Lair, Gustav Albrecht, Richard N. Lewis, and Sidney Goldstein.
7. U.S. Congress House of Representatives Committee on Un-American Activities, "Hearings Regarding Communist Infiltration of Radiation Laboratory and Atomic Bomb Project at the University of California, Berkeley, California," 81st Congress, April 22, April 26, May 25, June 10, June 14, 1949 (United States Government Printing Office, 1949).
8. Press release, May 27, 1949, Bohm file, Olwell archives, Mudd Library, Princeton University.
9. "Report on Soviet Espionage Activities in Connection with the Atomic Bomb (United States Government Printing Office, 1949).
10. B/W; Olwell, "Princeton, David Bohm and the Cold War"; Sam Schweber, e-mail correspondence, November 1993.
11. Interview with Rossi Lomanitz.
12. For information on the reactions of Princeton University and its staff, I am indebted to Russel Olwell for supplying me with his thesis and research papers.
13. Harold W. Dodds, "Communism and the Defense of America," occasional papers, the University of Hawaii, Honolulu, 1949.
14. Olwell reporting Schweber in Olwell, "Princeton, David Bohm and the Cold War."
15. Olwell referring to the David Bohm file in the Mudd Library, Princeton.
16. Interview with Rossi Lomanitz.
17. Olwell, "Princeton, David Bohm and the Cold War."
18. Martin Schwartzschild interviewed by Spencer Weart for the Bohr Library of the American Institute of Physics, July 19, 1979.
19. Olwell, "Princeton, David Bohm and the Cold War."

Chapter 7 Hidden Variables

1. David Pines, in *Quantum Implications: Essays in Honour of David Bohm,* B. J. Hiley and David Peat, eds., (London: Routledge, 1987).
2. Interview with Hanna Loewy, New York, April 1994.
3. Forty years later, the *Baltimore Sun* looked back on the events ("After 40 Years, Professor Bohm Re-emerges," *Baltimore Sun,* April 1990). Written by H. K. Fleming, a former managing editor, the article claimed that Oppenheimer's secretary had been instructed to keep Bohm away from Oppenheimer. She is said to have walked down the corridor, informing other secretaries and the receptionist at the Institute for Advanced Study that "David Bohm is not to see Dr. Oppenheimer. He is not to see him." I have not been able to authenticate this anecdote.

   While Einstein's invitation to Bohm was well known at Princeton, there appears to be no documented evidence for what subsequently occurred. Miriam Yevick was not at all surprised at Oppenheimer's reaction. At the time she was associated with the Emergency Civil Liberties Committee, an organization opposed to Senator McCarthy and his anti-Communist campaign. The committee wanted Einstein to speak in its support on the occasion of his seventy-fifth birthday. Yevick, acting as intermediary, spoke to Einstein's assistant who intimated that Einstein would address the group and support their fight.

   As the day approached, however, Einstein wrote to the committee indicating that he would not attend the function but would write them a letter instead. It appeared, according to Miriam Yevick (interviewed by the author in Leonia, New Jersey, April 1994), that Oppenheimer was worried that anti-McCarthy comments by Einstein would create problems for his institute and had said, "Einstein is creating trouble again." Friends were annoyed that, on Einstein's birthday, the physicist had met with Zionists and other political groups but avoided association with the Civil Liberties Committee.

   In an article, "Einstein and the Press," in *Physics Today* (August 1994), pp. 30–36, Abraham Pais gives a different version of this story. He refers to a report in the *New York Times* (March 15, 1954) that claims Einstein refused to accept flowers from the group, saying, "You may bring flowers to my door when the last witch hunter is sentenced, but not before."
4. Letter, Albert Einstein to P. M. S. Blackett, April 17, 1951.
5. Interview with Mort Weiss, Laguna Beach, California, April 1994.
6. Undated letter from Bohm to Hanna Loewy, presumably written in Florida in 1951.

7. Letter, Bohm to Miriam Yevick, January 7, 1952.
8. David Bohm, *Quantum Theory* (New York: Prentice-Hall, 1951).
9. B/W
10. Quoted in Paul Arthur Schlipp, ed., *Albert Einstein: Philosopher-Scientist* (Evanston, IL: Library of Living Philosophers, 1949).
11. Interview with Miriam Yevick, Leonia, New Jersey, April 20, 1993.
12. B/W
13. B/W
14. "A Suggested Interpretation of the Quantum Theory in Terms of Hidden Variables I," *Physical Review,* vol. 85 (1952), pp. 166–179; "A Suggested Interpretation of the Quantum Theory in Terms of Hidden Variables II," *Physical Review,* vol. 85 (1952), pp. 180–193.
15. Undated letter, Bohm to Einstein.
16. Letter, Einstein to Bohm, December 15, 1951.
17. Undated letter (probably December 1951) Bohm to Einstein.
18. Interview with Hanna Loewy, July 1993.

Chapter 8 Brazil: Into Exile

1. Interview with Hanna Loewy, New York, 1993; and B/W.

   On one occasion as I was driving the Bohms toward Newark airport, New Jersey, the sky began to blacken and we were enveloped by an intense storm with frequent lightning and sweeping rain. Bohm, taken back in time, remarked that the weather was exactly the same as the day he had left for Brazil.
2. Letter, Bohm to Hanna Loewy, October 1951.
3. B/W
4. Letter, Bohm to Einstein, November 1951.
5. Letter, Einstein to Bohm, December 5, 1951.
6. Letter, Bohm to Miriam Yevick, September 2, (presumably 1951).
7. Letter, Bohm to Miriam Yevick, November 1951.
8. Letter, Bohm to Miriam Yevick, n.d.
9. Letter, Bohm to Melba Phillips, n.d.
10. B/W
11. Letter, Bohm to Melba Phillips, written on the day his passport was confiscated.
12. Ibid.
13. Letter, Bohm to Yevick, n.d.
14. Letter, Bohm to Yevick, January 5, 1952.
15. Ibid.

16. During the conference, Bohm noted that Feynman was "beginning to develop a social conscience": he had been struck by the contrast between the opulence of the dinner put on for the scientists at a local country club and the extreme poverty outside. When the local mayor entered the room with great pomp and ceremony, Feynman walked out in anger. Undated letter (presumably late 1951 or early 1952), Bohm to Melba Phillips.
17. Letter, Bohm to Yevick, January 5, 1952.
18. Letter, Bohm to Loewy, December 10, 1951.
19. Letter, Bohm to Yevick, received May 8, 1952.
20. Bohm in informal discussion with the author during the mid-1980s.
21. Eugene Gross as recalled by Miriam Yevick, in an interview with the author, Leonia, N.J., April 1993.
22. Letter, Bohm to Yevick, November 20 (presumably 1951 or 1952).
23. Letter, Bohm to Loewy, December 1951.
24. Letter, Wolfgang Pauli to Bohm, December 3, 1951.
25. Wolfgang Pauli to Fierz in K. von Meyenn, W. Pauli, *Wissenschaft Briefwechsel, Vol. IV, Part 1: 1950–1952,* pp. 131–2.
26. Ibid.
27. "Reply to a Criticism of the Causal Re-interpretation of Quantum Theory," *Physical Review,* vol. 85 (1953), pp. 389–90L.
28. Letter, Bohm to Yevick, n.d.
29. Letter, Bohm to Yevick, January 5, 1952(?).
30. Letter, Paul Feyerabend to the author, July 9, 1993.
31. Ibid.
32. Letter, Feyerabend to the author, August 30, 1993.

    Never having heard of such a marked reaction on the part of Bohr to Bohm's work, I wrote to Feyerabend for confirmation. His second letter not only repeated but emphasized the effect that Bohm's *Physical Review* papers appeared to have had on Bohr.
33. Ibid.
34. Letter, Bohm to Yevick, January 9, 1952.
35. Ibid.
36. Letter, Rosenfeld to Bohm, May 20, 1952.
37. Ibid.
38. Ibid.
39. Letter, Bohm to Loewy, n.d.

    In the mid-1970s Rosenfeld repeated to me the position that, where quantum theory is concerned, there is nothing to argue about: "The phrase 'Copenhagen Interpretation' is actually a misnomer, in the sense that there is only one interpretation of quantum mechanism. . . . The misunderstandings that have been expressed so vociferously from various sides are based on a disregard of this circumstance. They take the formalism and then they try to put upon it what they call an interpretation, without reflecting that the way in which the equations are written already implies a definite interpretation, that is, a definite relationship between the symbols and physical concepts." See Paul Buckley and F. David Peat, *A Question of Physics* (Toronto and Buffalo: University of Toronto Press, 1979). The updated and revised version of this book is entitled *Glimpsing Reality: Ideas in Physics and the Link to Biology* (Toronto, Buffalo and London: University of Toronto Press, 1996).

40. Letter, Bohm to Yevick, March 9, 1952.
41. Letter, Bohm to Yevick, January 9, 1951.
42. Letter, Bohm to Yevick, January 13, 1952.
43. Letter, Bohm to Yevick, n.d.
44. Letter, Bohm to Yevick, February 16 (year uncertain).
45. Ibid. A similar letter was written to Melba Phillips.
46. Ibid.
47. Ibid.
48. Letter, Einstein to Max Born, May 12, 1952.
49. Letter, Einstein to Born, October 12, 1953.
50. Letter, Einstein to Likpin, July 5, 1952.
51. Max Dresden, remarks from the floor at the American Physical Society Meeting, Washington, May 1989. Dresden confirmed this version in an interview with the author immediately following that session and in a letter to the author.
52. Letter, Bohm to Yevick, n.d.
53. Ellen W. Schrecker, *No Ivory Tower: McCarthyism and the Universities* (New York: Oxford University Press, 1986). See also Walter Goodman *The Committee: The Extraordinary Career of the House Committee on Unamerican Activities,* (New York: Farrar, Straus and Giroux, 1964) and Harold W. Dodds, *Communism and the Defense of America,* occasional papers, University of Hawaii, Honolulu, Hawaii, 1949.
54. Ellen W. Schrecker, interviewed by the author, New York, August 1993.
55. Peter Bergman, interviewed by the author, New York, August 1993.
56. Melba Phillips, interviewed by the author, New York, July 1994.
57. Letter, Feyerabend to the author, August 30, 1993.
58. Letter, Bohm to Phillips, January 9, 1952.

59. Letter, Bohm to Yevick, February 28, 1952.
60. Letter, Bohm to Yevick, January 13 (1952 or 1953).
61. Letter, Bohm to Yevick, March 15, 1954.
62. Letter, Bohm to Yevick, November 1951(?).
63. Letter, Bohm to Yevick, November 1951.
64. Letter, Bohm to Yevick, November 20, 1951.
65. Ibid.
66. Letter, Bohm to Loewy, 1952(?).
67. Letter, Bohm to Loewy, n.d.
68. Letter, Bohm to Loewy, February (year uncertain).
69. Letter, Bohm to Phillips, January 15, 1952.
70. Ibid.
71. Ibid.
72. Ibid.
73. Letter, Bohm to Yevick, January 2, 1952.
74. Letter, Bohm to Yevick, n.d.
75. Letter, Bohm to Yevick, March 17, 1952.
76. Ibid.
77. Letter, Bohm to Yevick, October 24, 1953.
78. Letter, Bohm to Yevick, November 22, 1952.
79. Letter, Bohm to Yevick, January 1954.
80. Letter, Bohm to Yevick, November 20, 1951.
81. Ibid.

CHAPTER 9 CAUSALITY AND CHANCE

1. Letter, Bohm to Miriam Yevick, May 8, 1952.
2. Ibid. A similar (undated) letter was written to Melba Phillips.
3. Ibid.
4. Letter, Bohm to Yevick, May 12, 1952.
5. Letter, Abrahão de Moraes to Einstein, May 12, 1952.
6. Einstein's letter, written in response to de Moraes's request, May 24, 1952.
7. From a second letter from Einstein, in response to de Moraes, written on the same day, May 24, 1952.
8. Letter, Bohm to Melba Phillips, 1951.
9. Letter, Bohm to Phillips, June 28, 1952.
10. Letter, Bohm to Yevick, October 16, 1952.
11. George Yevick interviewed by the author in Ottawa, 1994.
12. Vigier interviewed by the author in Brussels, 1995.

13. Letter, Bohm to Yevick, December 24 (year uncertain).
14. Letter, Bohm to Yevick, n.d.
15. Ibid.
16. Letter, Bohm to Yevick, October 16, 1952(?).
17. Letter, Bohm to Yevick, n.d.
18. Letter, Bohm to Yevick, August 19, 1952.
19. Letter, Bohm to Yevick, n.d.
20. Letter, Bohm to Yevick, October 24, 1953.
21. Letter, Bohm to Yevick, n.d. (included among a package of letters for 1954).
22. Letter, Bohm to Yevick, January 1954.
23. Letter, Bohm to Yevick, April 2, 1953.
24. Letter, Bohm to Yevick, December 1951(?).
25. Letter, Bohm to Yevick, February 15, 1952.
26. Letter, Bohm to Yevick, n.d.
27. Letter, Einstein to Bohm, January 22, 1954.
28. Letter, Einstein to Bohm, February 10, 1954.
29. Letter, Einstein to Bohm, October 28, 1954.
30. Ibid.
31. Letter, Bohm to Einstein, November 14, 1954.
32. Letter, Einstein to Bohm, November 24, 1954.
33. Ibid.
34. Letter, Bohm to Yevick, n.d.
35. Letter, Bohm to Oppenheimer, November 29, 1966.
36. Letter, Bohm to Yevick, n.d.
37. Ibid.
38. Ibid.
39. Letter, Bohm to Yevick (late summer of 1954).
40. Letter, Bohm to Yevick, September 10, 1954.
41. Letter, Bohm to Yevick, November 5, 1954.
42. Letter, Bohm to Melba Phillips, March 18, 1955.
43. Letter, Bohm to Yevick, Spring 1955.

Chapter 10 Israel: The World Falls Apart

1. Letter, Bohm to Miriam Yevick, Spring 1955.
2. Interview with Saral Bohm.
3. Ibid.
4. Ibid.
5. Letter, Bohm to Yevick, Spring 1955.

6. Interview with Miriam Yevick, April 30, 1993.
7. Interview with Saral Bohm, January 1995.
8. Interview with Yevick, April 30, 1993.
9. Interviews with Basil Hiley, February 1996.
10. Interview with Yevick, April 30, 1993.
11. Interview with Saral Bohm, May 1993.
12. Ibid.
13. Interview with Saral Bohm.
14. Ibid.

CHAPTER 11 BRISTOL: ENCOUNTERS WITH FAMOUS MEN

1. Paul Buckley and F. David Peat (eds.), *A Question of Physics* (Toronto and Buffalo: University of Toronto Press, 1979). The updated and revised version of this book is entitled *Glimpsing Reality: Ideas in Physics and the Link to Biology*, Toronto, Buffalo and London: University of Toronto Press, 1996).
2. Interview with Maurice Wilkins, London, March 1994.
3. B/W
4. B/W
5. Bohm interviewed by Evelyn Blau, Brockwood Park, England, September 2, 1978.
6. Interviews with Saral Bohm.
7. Letter, Paul Feyerabend to the author, July 9, 1993.
8. Interviews with Saral Bohm.
9. Ibid.
10. David Bohm, *Causality and Chance in Modern Physics* (Princeton, N.J.: Van Nostrand, 1957).
11. Letter, Robert G. Chambers to the author, April 13, 1994.
12. Interview with Basil Hiley, London, February 1996.
13. Ibid.
14. The following outline of Krishnamurti's life was gleaned from interviews listed individually and these biographies: Mary Lutyens, *Krishnamurti, The Years of Awakening* (London: John Murray, 1975); Mary Lutyens, *Krishnamurti: The Years of Fulfillment* (London: John Murray, 1983); Pupul Jayakar, *Krishnamurti: A Biography* (San Francisco: Harper and Row, 1986); and Radha Rajagopal Sloss, *Lives in the Shadow with J. Krishnamurti* (London: Bloomsbury and Reading, MA: Addison-Wesley, 1991).
15. Interviews with Saral Bohm and earlier informal conversations with David Bohm.

16. David Pines and David Bohm, "A Collective Description of Electron Interactions: II. Collective vs. Individual Particle Aspects of Interactions," *Physical Review,* vol. 85 (1952), pp. 338–353. The two other co-authored papers in this series were: Bohm and Pines, "A Collective Description of Electron Interactions: I. Magnetic Interactions," *Physical Review,* vol. 82 (1951), pp. 625–634 and Bohm and Pines, "A Collective Description of Electron Interactions: III. Coulomb Interactions in a Degenerative Electron Gas," *Physical Review,* vol. 92 (1953), pp. 609–625.
17. Pines preserved his 1952–55 correspondence with Bohm, believing it was of historical interest. But the letters were destroyed in a fire in the early 1980s. Letter, Pines to author, October 3, 1994.
18. Thomas Kuhn, *The Structure of Scientific Revolutions* (Chicago: University of Chicago Press, 1962).
19. B/W; and interview with Gudeon, London, February 1994.
20. Telephone interviews with Irving Bohm during 1994.
21. Werner Heisenberg, *Physics and Philosophy: The Revolution in Modern Science* (New York: Harper, 1958).
22. David Bohm, "Classical and Non-Classical Concepts in Quantum Theory," *British Journal for the Philosophy of Science,* vol. 12, no. 48 (February 1962), pp. 265–280.
23. Interview with Basil Hiley, London, January 18, 1995.
24. This would have been the English translation of de Broglie's *Une Tentative d'interprétation causale et non linéaire de la mécanique ondulatoire,* (Paris: Gautier-Villars, 1956).
25. B/W

Chapter 12 Birkbeck: Thought and What May Lie Beyond

1. Interviews with Saral Bohm.
2. B/W
3. Interview with Paavo Pylkkanen, January 12, 1995.
4. Interview with J. P. Vigier, Brussels, March 1995.
5. Interview with Basil Hiley; and B/W.
6. Interview with Vigier.
7. Mark Edwards and David Bohm, *Changing Consciousness: Exploring the Hidden Sources of the Social, Political, and Environmental Crises Facing Our World* (San Francisco: Harper San Francisco, 1991).
8. Interviews with Mary Cadagon, London, February and March 1994; interviews with Saral Bohm.

9. Bohm in conversation with the author, Ottawa, 1975.
10. Interviews with Saral Bohm.
11. Interviews with Mary Cadagon.
12. Interviews with Saral Bohm.
13. Interviews with Mary Cadagon.
14. Bohm in conversation with the author, Ottawa, 1974.
15. Interviews with Mary Cadagon.
16. Ibid; and interviews with Saral Bohm.
17. Bohm, during several conversations with the author in the late 1980s.
18. Charles Biederman, *The New Cézanne: From Monet to Mondrian* (Red Wing, MN: Art History, 1958); and Charles Biederman, *Art as the Evolution of Visual Knowledge* (Red Wing, MN: Art History, 1948).
19. Interview with Antony Hill, London, March 1994.
20. Letter, Bohm to Biederman, April 24, 1960.
21. Letter, Biederman to Bohm, May 22, 1960.
22. Alfred Korzybski, *Science and Sanity: An Introduction to Non-Aristotelian Systems and General Semantics,* 4th ed. (orig. 1933; Lakeville, CT: International Non-Aristotelian Library Publishing Co., 1958).
23. Letter, Bohm to Biederman, June 6, 1960.
24. Letter, Bohm to Biederman, December 31, 1962.
25. Letter, Bohm to Biederman, April 24, 1960.
26. Letter, Biederman to Bohm, May 22, 1960.
27. Letter, Bohm to Biederman, September 23, 1960.
28. W. V. D. Hodge, *The Theory and Application of Harmonic Integrals* (New York: Cambridge University Press, 1959).

Chapter 13 Language and Perception

1. Interviews with Basil Hiley.
2. Ibid.
3. David Bohm, *The Special Theory of Relativity* (New York: W. A. Benjamin, 1965).
4. Referee's report to Robert R. Worth, editorial director, Benjamin (publisher), November 27, 1963.
5. Telephone interview with Schumacher, 1994.
6. Sir Arthur Conan Doyle, "The Final Solution," *Memoirs of Sherlock Holmes* (Oxford University Press, 1993).
7. Interviews with Saral Bohm.
8. Telephone interview with Schumacher.
9. Interview with Joe Zorskie, Ojai, April 1994.

10. An account of this meeting between Western physics and Native American philosophy can be found in F. David Peat, *Lighting the Seventh Fire* (New York: Carol Publishing Group, 1994).
11. Telephone interview with Schumacher.
12. Interview with Saral Bohm, May 1993.
13. Telephone interviews with Schumacher.
14. Interviews with Hiley.
15. Letter, Colgate to Keeny, April 19, 1965.
16. Letter, Keeny to Colgate, April 29, 1965.
17. Letter, Bohm to Lomanitz, November 21, 1966.

Chapter 14 The Implicate Order

1. David Bohm, *Wholeness and the Implicate Order* (London and Boston: Routledge and Kegan Paul, 1980).
2. Letter from Karl Pribram to the author.
3. Interview with David Hockney, London, November 1995.
4. Ian McEwan, *The Child in Time* (Boston: Houghton-Mifflin, 1987).
5. Telephone interview with Ian McEwan, March 1995.
6. Ibid.
7. Interview with Christopher Philippidis, Bristol, 1995.
8. Interviews with Basil Hiley.
9. David Bohm and Basil Hiley, *The Undivided Universe* (London: Routledge, 1993).
10. Interviews with Hiley.
11. Interview with Joe Zorskie, Ojai, California, April 1994.
12. Interview with Hiley, February 1996.
13. Interview with David Moody, Ojai, April 1994.
14. Interview with Booth Harris, Ojai, April 1994; and interviews with Saral Bohm.
15. Interview with Booth Harris.
16. Interview with Lee Nichol, Ojai, California, April 1994.
17. Ibid.
18. Interview with Mort Weiss, Laguna Beach, California, April 1994, and subsequent telephone conversations and letters.
19. Interviews with Saral Bohm and Basil Hiley.
20. Interviews with Hiley.
21. Owen Barfield, *Poetic Diction: A Study in Meaning* (London: Faber and Faber, 1952); and Owen Barfield, *Speaker's Meaning* (Middletown, CT: Wesleyan University Press, 1967).

22. David Bohm and F. David Peat, *Science, Order and Creativity* (Toronto and New York: Bantam Books, 1987).
23. Bohm to the author, during an informal conversation the day following this dinner.
24. Interviews with Saral Bohm.
25. Part of this work later appeared in "A New Theory of the Relationship of Mind and Matter," *Phil. Psych.*, vol. 3 (1990), pp. 271–286; and "Eine Neue Theorie zur Beziehung Zwischen Geist und Materie" in *Wissenschafter und Weise: Die Konferenz,* Petra Michel, ed. (Grafing: Aquamarin Verlag, 1991), also published as "A New Theory of the Relationship of Mind and Matter," *J. Am. Society for Physical Research,* vol. 80 (1986), pp. 113–135.
26. Interview with Rupert Sheldrake, London, January 1995.
27. Alex Comfort, *The Joy of Sex* (New York: Crown, 1972); and Alex Comfort, *Reality and Empathy: Physics, Mind and Science in the Twenty-first Century* (Albany: State University of New York Press, 1984).

Chapter 15 Dialogue and Disorder

1. Interview with Edward Gudeon, 1994.
2. Papers and brief held by Edward Gudeon, London.
3. Interviews with Catherine Shainberg, 1994.
4. Interview with Harvey Brown, Oxford, 1995.
5. Interviews with Saral Bohm, and earlier informal conversations with David Bohm.
6. Information in this section was gathered during interviews with Booth Harris, Mark Lee, David Moody, Lee Nichol, Mary Zimbalist, and Joe Zorskie at Ojai, California, April 1994. Additional information was supplied by Scott Forbes and Harsh Tanka during interviews at Brockwood Park, England, in February 1994.
7. Ibid.
8. Interview with Paul Grof, July 1994.
9. Interview with Patrick de Mare, London, March 1994.
10. Suzi Gablik, *Conversations Before the End of Time* (New York: Thames and Hudson, 1995).
11. Toward the end of his life Bohm spoke increasingly about dialogue. Reference to the process is made in David Bohm and F. David Peat, *Science, Order and Creativity* (Toronto and New York: Bantam Books, 1987); *Unfolding Meaning: A Weekend of Dialogue with David Bohm,*

Donald Factor, ed. (Mickleton, England: Foundation House Publishers, 1985); and the posthumous *Thought as a System,* Lee Nichol, ed. (London: Routledge, 1994).

12. Interviews with Don Factor and Peter Garret, London, March 1994.
13. Interview with Joan Evans, London, March 1994.
14. Informal discussions with David Shainberg during the early 1990s; and interviews with Factor, Garret, and Evans.
15. The author was a participant at this meeting.
16. The author was a participant at these meetings.
17. Informal discussions with David Bohm during the 1980s. The proceedings of this meeting were published in *Towards a Theoretical Biology, Vol. 2,* C. H. Waddington, ed. (Chicago: Aldine, 1970).
18. Interview with Roger Penrose, Oxford, March 1994.
19. Interview with Joe Zorskie, Ojai, April 1994.
20. Interviews with Lee Nichol, Ojai, California, April 1994.
21. Interview with David Moody, Ojai, California, April 1994.
22. Interview with Joe Zorskie, Ojai, California, April 1994.
23. I was present when Karl Pribram questioned J. Krishnamurti during the first "Dialogue with Scientists" meeting at Brockwood Park in April, 1974.
24. Bohm interviewed by Evelyn Blau, Brockwood Park, England, September 2, 1978.
25. Interview with Paavo Pylkkanen, London, January 23, 1995.
26. Saral Bohm interviews. David Bohm related this story to me on several occasions.

Chapter 16 The Edge of Something Unknown

1. David Bohm, informal conversations with the author; interviews with Saral Bohm; interview with Lee Nichol, Ojai, California, April 1994; interview with Don Factor, London, March 1994; interviews with Basil Hiley.
2. Radha Rajagopal Sloss, *Lives in the Shadow with J. Krishnamurti* (London: Bloomsbury, 1991 and Reading, MA: Addison-Wesley, 1993).
3. Interview with David Moody, Ojai, California, April 1994.
4. Telephone conversations between David Bohm and the author during 1991; interviews with Basil Hiley and Saral Bohm during February and March 1994.
5. Interviews with Saral Bohm.

6. Interview with Carolyn Shainberg, New York, April 1994.
7. Interview with Patrick de Mare, London, March 1994; and interviews with Saral Bohm.
8. Interviews with Saral Bohm.
9. Letter from Dr. Levy to the author, March 17, 1994, and telephone interview, March 15, 1994.
10. Ibid.
11. Interview with Booth Harris, Ojai, California, April 1994.
12. Letter from Dr. Levy, and telephone interview.
13. Interview with Paavo Pylkkanen, London, January 1995; interview with Don Factor, London, April 1994.
14. Interview with Joan Evans, London, March 1994.
15. Telephone interview with Lee Nichol and Saral Bohm, April 19, 1994; subsequent discussion with Lee Nichol in Ojai, April 1994.
16. Interview with Saral Bohm, February and March 1994.
17. Interviews with Saral Bohm; interview with Carolyn Shainberg, New York, April 1996; interview with Paul Grof, Ottawa, May 1994.
18. Letter from Levy to author, March 17, 1994.
19. Interview with Saral Bohm, London, March 1994; letter from Levy.
20. Interviews with Basil Hiley.
21. Ibid.
22. Ibid.
23. David Bohm during several conversations with the author at Kalamazoo during April 1992; and interviews with Saral Bohm in London during January and February of 1994.
24. Interview with Lee Nichol and with other dialogue participants, Ojai, April 1994.
25. Informal discussions with Paul Grof, Ottawa, 1994; letter, Levy to Grof.
26. Interview with Therese Schroder-Sheker, Kalamazoo, April 1992.
27. E-mail communications with Ivan Havel.
28. Interviews with Saral Bohm during February and March 1994, and January 1995, London.
29. Interview with Pylkkanen, London, January 1995.
30. Interviews with Nichol, Ojai, April 1994.
31. Interview with Factor, London, March 1994.
32. Interviews with Saral Bohm, London, February and March 1994.
33. Interviews with Basil Hiley, London, February and March 1994; and interviews with Saral Bohm, February and March 1994. Understanding how easily memories can be transformed by time, Saral immediately

wrote down an account of everything that had occurred on her husband's last day.

34. David Bohm and Basil Hiley, *The Undivided Universe: An Ontological Interpretation of Quantum Theory* (New York and London: Routledge, 1993).
35. Henry Stapp, review of *The Undivided Universe* in *American Journal of Physics,* vol. 62 (1994).
36. Daniel M. Greenberger, "Interpretations," *Science,* vol. 266 (1994), p. 147.
37. Sheldon Goldstein, review of *The Undivided Universe* in *Physics Today* (September 1994), p. 90.
38. Chris Clarke, review of *The Undivided Universe* in *Network,* no. 54 (Spring 1994), pp. 42–44.
39. Lee Nichol, ed., *Thought as a System* (London: Routledge, 1994).

In addition to the sources referenced above, the author was able to obtain general insights and background information on David Bohm's life and work from the following generous people:

John Briggs, telephone conversations with the author, over many years.

Keith Critchlow, interview by the author, March 1994.

Mark Edwards, numerous conversations with the author, after February 1994.

Mark Lee, interview by the author, April 1994.

Yuval Portugali, letter to the author, January 1994.

David Schindler, interview by the author, October 1993.

David Schrum, many informal conversations.

David Shainberg, many discussions before his premature death. Shainberg, a painter and psychotherapist, had been a close friend of the Bohms for many years. David Shainberg had a number of conversations with David Bohm about his childhood and psychology. We had planned to carry out some focused discussion for this biography, but the day I arrived in New York to interview him, David Shainberg lapsed into a coma from which he did not recover.

INDEX